SUSTAINABILITY OF SMALL HOLDER AGRICULTURE IN INIDA

About the Author

Dr. T. Haque is a well known Agricultural Economist, having specialisation in the field of Agricultural Development and Policy. He obtained his Ph.D. degree from Patna University and Post-Doctoral training in Agricultural Development Policy from Iowa State University, Ames, USA. Recently, Dr. Haque has been offered an award of ICAR National Fellow and he is working at the National Centre for Agriculture Economics and Policy Research, New Delhi. Prior to this, he served the National Institute of Rural Development, Hyderabad as Director, Centre for Agrarian Studies for over five years. Between 1976 and 1989, he served as Associate Professor/Assistant Professor (Scientist) at the Indian Agricultural Research Institute, New Delhi.

Dr. Haque served as Consultant/Resource Person of several National and International Organisations including FAO and ILO. Recently during March-August 1994, he served as a Member of FAO Mission to help the Government of Vietnam in restructuring the agrarian institutions in Vietnam.

Dr. Haque is an author of several books including : (i) Agrarian Reforms and Institutional Changes in India; (ii) Impact of Land Reform on Agriculture and Rural Development; (iii) Transfer of Technology to Small Farmers. Besides he has published about 70 Research papers and some of his research papers have been widely quoted/referred by eminent economists and other social scientists.

SUSTAINABILITY OF SMALL HOLDER AGRICULTURE IN INDIA

By

T. HAQUE

CONCEPT PUBLISHING COMPANY, NEW DELHI-110059

ISBN 81-7022-578-7

First Published 1996

Published and Printed by
Ashok Kumar Mittal
Concept Publishing Company
A/15-16, Commercial Block, Mohan Garden
NEW DELHI 110059 (India)
Phones: 5554042, 5504042

To

My Daughter

HINA

FOREWORD

Due to growing pressure of population on land, the average size of holding in India is gradually declining overtime. At present, small and marginal farms account for nearly 78 per cent of the total farm households in the country. Small size of farm by itself may not be a constraint to agricultural development. Countries like Japan and Korea have managed to achieve high levels of agricultural development with very small sizes of holdings. What is needed is appropriate infrastructural, institutional, technological, organisational and policy support for the sustainable development of small farms and their integration with non-farm development. In the long run, a substantial proportion of population may have to shift from agriculture to non-agricultural sector. But in the short run, we need to evolve a method by which a vast majority of India's agricultural population could be made viable.

The book entitled 'Sustainability of Small Holder Agriculture' by Dr. T. Haque is an excellent treatise on the subject. This provides a comprehensive analytical framework as to how our small and marginal farms could be helped to sustain and join the mainstream of development through appropriate agricultural support service system and proper integration of farm and non-farm sectors, particularly in the wake of new economic policy of liberalisation and market reform.

Besides, the book is based on an All India Research Study and provides enormous data on factor productivity of agricultural enterprises including crop production, horticulture, animal husbandry, fishing etc., in various states. The book also analyses the questions of land tenure and institutional arrangements of agriculture at great-length. I am sure, the book will be of great value and interest to policy-makers, planners, agricultural scientists, students and all those who are concerned about accelerated development of small farm dominated agricultural economy of India.

N.S. RANDHAWA
Former Director-General I.C.A.R.

ACKNOWLEDGEMENTS

The book entitled 'Sustainability of Small Holder Agriculture in India', is based on an All India Research Study completed by me at NIRD recently. I am extremely thankful to Shri T.L. Sankar, Former Director-General of NIRD, Sri A.V.S. Reddy former Deputy Director-General, NIRD and Sri P. Vanamali, Present Deputy Director-General, NIRD for their constant support and encouragement. Besides, I am grateful to Prof. V.S. Vyas, Prof. Y.K. Alagh, Prof. N.S. Randhawa, Prof. R.S. Paroda, Prof. S.S. Johl, Prof. C.H. Hanumantha Rao, Sri K.B. Saxena, Dr. C.C. Maji, Dr. D. Jha, Dr. M.S. Bhatia, Dr. C. Ramaswamy, Prof. I.J. Singh, Dr. Mruthyunjaya, Prof. Radha Krishna, Prof. G. Parthasarathy, Prof. Narsimha Reddy, Dr. P.C. Bansie, Mr. L. Montesi, Mr. D.P.A. Naidu, Dr. R.P. Singh and Dr. Rajagopal for their moral support and encouragement. I also wish to place on record the research assistance provided by Adarsh Bala Sachdeva, Shyam Sunder, Riyaz, Sushma, Satbir Sirohi and Pratap Singh. However, I take responsibility for errors if any.

T. HAQUE

CONTENTS

LIST OF TABLES

LIST OF FIGURES

1

INTRODUCTION

Today farmers in India are found to be in a depressed, albeit angry mood. The depression is said to have resulted partly from growing rural-urban disparities and partly from declining trends in profitability of agriculture in almost all the states, the rates of decline being higher in the more developed states (Haque, 1992). The farmers are also angry under the impression that the Government under the direction of World Trade Organization (WTO) may withdraw various forms of farm protection which may affect their future prospects adversely. While the reasons for relatively greater decline in agricultural profitability in developed states which are the main beneficiaries of green revolution would need to be explored, the declining trend in profitability in the backward regions is a matter of more serious concern because the levels of technology adoption and consequently the yields and income levels of farmers in backward regions are awfully low. This raises the basic question of viability and sustainability of agriculture dominated by small and marginal farms that do not produce enough even for subsistence not to speak of generating surplus for investment and growth.

Small and marginal holdings account for nearly 78 per cent of the total operational holdings in the country, operating over 32 per cent of the total area. Given the present trends and sectoral patterns of population growth and labour absorption, it is obvious that the proportion of small and marginal farms in the country would continue to grow, and therefore, it is no use merely saying that our small and marginal farms are non-viable and non-sustainable. The primary question to-day is how to make them viable and sustainable or alternatively to provide adequate employment opportunities for them in the non-farm sector. In the wake of new economic policy of liberalisation, it is often anticipated as well as apprehended that India's farm sector would undergo a process of rapid capitalistic change, thereby leading to concentration of land holding in the hands of a few and proletarisation and pauperisation of the rural masses. This is particularly true if we withdraw ceiling on land holdings or allow the entry of multinationals or domestic corporate bodies are allowed to operate without restrictions. But the question is : can we prepare ourselves for such a change without exploring adequate employment opportunities for the masses or without going for drastic cut on population growth? As Dantwala (1992) has rightly pointed out, the most difficult problem confronting Indian Agriculture is to evolve policies and operational systems to make the small farm economy efficient so that it may provide not only a decent income to those who depend on it for their livelihood, but also meet the national requirement of food and raw materials. In the long run, a substantial proportion of agricultural population may have to go for non-agricultural occupation which may ease the situation. But in the short run, we are confronting a situation in which majority of our small and marginal farms are said to be non-viable and therefore, we may have to evolve suitable strategies and policy options for making them viable and sustainable.

A farm is said to be viable when it generates sufficient income to maintain a family at reasonable level by employing the family labour effectively. If such viability is assured on regular basis, then the farms can also be called sustainable. Some recent studies (Maji *et al.* 1990; Acharya and Verma, 1990) from various regions show that although our marginal and small farms are as efficient as the large farms (if not more), they do not generate sufficient income to sustain the land holder's family. This is particularly true in the case of marginal farmers in dry, albeit backward regions where they depend greatly on wage employment for their subsistence. The situation is chaotic where the wage employment opportunities are low. Since marginal farmers are net buyers of foodgrains, the rising prices of foodgrains also affect them adversely. This point needs special attention particularly when our policy-makers seem to depend more on price and market mechanism than on infrastructural and institutional facilities for agricultural development.

There is no denying the fact that there is tremendous potential for increasing yield and income which remains largely untapped for lack of suitable infra-structural and institutional facilities as well as appropriate technology cum policy measures (Govt. of India, Report of the High Powered Committee on Agricultural Policies and Programmes, 1990; Swaminathan, 1986).

Particularly, there is need to examine and suggest measures for improved adoption of new production technology by small and marginal farmers and gradual commercialisation of farms. What make the small and marginal farms to become low or slow adopters of any new production technology? Will the current market reform programme help adopt new technology by small farms or alternatively do we need to further strengthen government driven and other institutional facilities of credit, marketing and irrigation which would enable the small and marginal farms to generate sufficient output and income for their sustenance? What would be the role of low cost bio-technology or bio-fertilizers in this regard? These are some of the questions which remain unanswered, particularly in the context of agriculturally backward regions of the country.

The experience of countries like Japan and Korea where the operational sizes of holdings are relatively very small, shows that although the level of agricultural efficiency is quite high, majority of the small farmers do not generate sufficient income from farm alone for their sustenance. There are other supplementary activities which help them grow (Kaneda, 1980). Besides, the additional growth of labour force in the rural sector has been invariably absorbed in the non-agricultural sector thereby disallowing the overcrowding of population in agriculture. Unfortunately, both these elements are weak in Indian context. Therefore, while recognising the relevance of small scale Japanese agriculture as a model for India because of its low land—man ratio and limited employment opportunities outside of agriculture on one hand and possibilities of raising yields on small farms on the other, we may have to look for rapid growth of rural diversification and absorption of additional labour force outside of agriculture.

Economists have been studying and debating the role of size of farm in agricultural development for the past several decades. However, it has not yet been established that small size of land holding acts as a constraint on rapid agricultural development, particularly in the light of Japanese and Korean experiences of agricultural development with appropriate technology and farm organisations.

While most of the pre-green revolution period (Raj Krishna, 1962; Sen, 1962; Hopper, 1965) studies confirmed inverse relationship between size of farm and productivity, some of the post-green revolution studies (Parthasarathy, 1987; C.H.H. Rao, 1975) show that inverse relationship under traditional technology does not seem to hold true when new technologies are introduced as large farms invest more in modern inputs per unit of area and productivity shows no significant relationship to farm size, despite higher irrigation and cropping intensities on small farms. Since small farmers have relatively less access to modern inputs including HYV seeds, chemical fertilizers, pesticides, information and credit and that they are consequently slow to adopt new technology, the inverse relationship between farm size and output per acre is not only being weakened, but in some cases reversed. However, WCARRD Report (FAO, 1989) documented studies which show that in greater parts of Asia and the Pacific, "the rate of adoption of new technologies especially those based on non-monetary and low monetary inputs by the small farms were as high as the large farmers.... Even if small farmers are slow to start with, they evidently catch up and finally use the modern technologies as much as or even more efficiently than the large farmers." Furthermore, Binswanger and Rosenweig observed that operated area continues to be negatively associated with land productivity, while owned area is positively correlated with land productivity. In short, the controversy regarding the size-productivity relationship still remains and one needs to examine in greater details the underlying reasons for such controversy. This would however, require more representative samples covering various agro-climatic regions and more appropriate methods of analysis than what have been employed by the erstwhile scholars.

Specially, the issue needs to be examined in the light of the following questions :

1. What are the possible factors that influence size—productivity relationship?
2. Do the differences in land quality, management, access to modern inputs, factor—product prices and the differential socio-economic goals and incentives of two sets of farms matter most or size by itself has any role to play?
3. Can small farms contribute to much talked about commercialization/industrialisation of agriculture by freeing themselves from traditional subsistance trap and proper integration with local and global markets? Can small farms compete with large corporate farms or multinationals in the production of export oriented agricultural commodities?

Answers to these questions will largely help end the controversy. Therefore, one of the objectives of this study is to provide answers to some of these basic questions relating to size-productivity relationship.

Similarly, although the negative role of share cropping or for that matter all other forms of tenancy *vis-a-vis* owner operated farms has been widely recognised, (Heady, 1947; Bardhan and Srinivasan, 1971, and C.H.H. Rao, 1971), there are studies (Cheung 1968; Reid, 1975) which indicate that there is no relationship between tenurial system and farm productivity, as the landlords can effectively and inexpensively monitor tenants' effort on share cropped land. Besides, in the relatively developed regions of India, a system of commercial tenancy (reverse tenancy) has emerged which is said to be as efficient as the owner cultivation. Whatever be the case, there is reasonable scope for further investigation into the exact nature of relationship between tenurial status on the one hand and farm productivity and income on the other.

Moreover, even if we assume that there is limit beyond which agriculture cannot absorb and sustain such a huge proportion of small and marginal farmers as well as landless labourers and that there is need for rapid transfer of population from agriculture to non-agriculture, our policy goal should be made far more clear whether we can influence migration more effectively by coercing the landless and marginal farmers who do not have necessary skills and resources to take up new things more effectively (who in all probability would contribute to urban slums only) or pressurise the rural rich to migrate who have relatively greater access to human as well as financial resources which are so necessary for any efficient entry into new areas. As IP, PC and CS Stahl (1978) point out, entrepreneurship is the most scarcest factor of production in developing countries and that entrepreneurial functions are more demanding in industry, government and commerce than in agriculture and that landlords are usually better educated and better trained than the tenants and hence possess more entrepreneurial talent. The land lords's class' often very high levels of income have placed them on the backward bending portion of their supply of effort function. With land redistribution, there will occur wealth redistribution which would impel the former landlord class to rely for income creation on their entrepreneurial abilities rather than their wealth.

Drawing on the experience of Japanese development, W.A. Lewis (1955) observed that a Swift transformation of land owners and nobles into capitalists...resulted from the state buying out the feudal rights of the nobility and depriving them of their administrative functions; at the same time, it also took over the debts of the feudal lords. Finding themselves with plenty of money (or rather government bonds) and no duties, some of the lords turned in the first place to banking and when the Government decided that it was ready to sell some of the factories which it had established for pioneering purposes, it found a ready market. This Swift transformation of an old type aristocracy into a new type capitalist class was profoundly important in increasing the supply of entrepreneurship in Japan in the crucial last quarter of the 19th century. Furthermore, land reform which transfers the ownership of land to the tiller of the soil, is likely to generate the greatest amount of efficiency, because the new owner cultivator has all the incentive and motivation to do his best. (Leibenstein, 1966). Thus, land redistribution can augment society's stock of entrepreneurial talent. Although, there has not been any empirical investigation into the relationship between land reform and non-farm development in the past, the identification of the nature, extent and causes of migration of various classes of population in the rural areas in the course of our study would throw some light on this issue. But again this question has to be examined against the background of new economic policy of liberalisation and market reform.

Objectives of the Study

Keeping in view the above mentioned facts, the present study was undertaken with the following main objectives :

(i) to analyse the profitability of various agricultural enterprises including crop production, horticulture, animal husbandry. fishing etc.;

(ii) to examine the viability and sustainability of small and marginal farms in the context of new economic policy of liberalisation and market reform;

(iii) to analyse the relationship between size of farm and tenurial status on the one hand and productivity/commercialisation of agriculture on the other; and

(iv) to indicate appropriate policy measures for viability and sustainability of small and marginal farms in India.

Data and Method

For the purpose of this study, both secondary and primary data were collected and analysed. While the secondary data were collected from the Directorate of Economics and Statistics, Ministry of Agriculture, Government of India and analysed with a view to examining the recent trends in the profitability of crop cultivation in various regions, these concealed many micro-level realities and also did not provide information

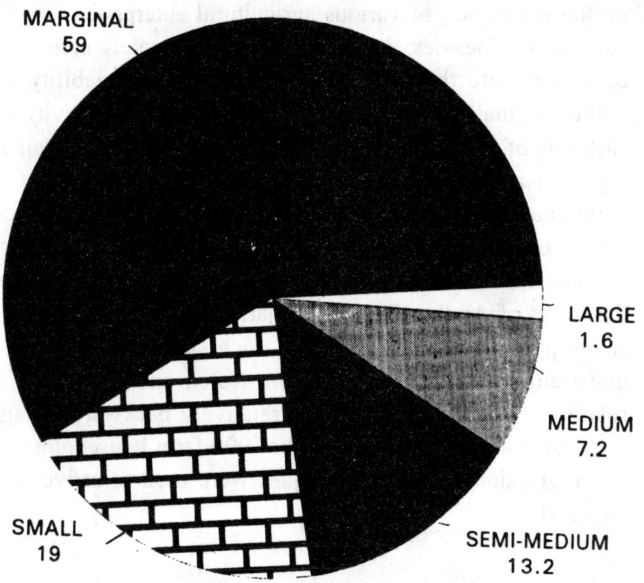

FIG. 1.1 : Percentage Distribution by Number of Operational Holdings by Size Group (as of 1990-91)

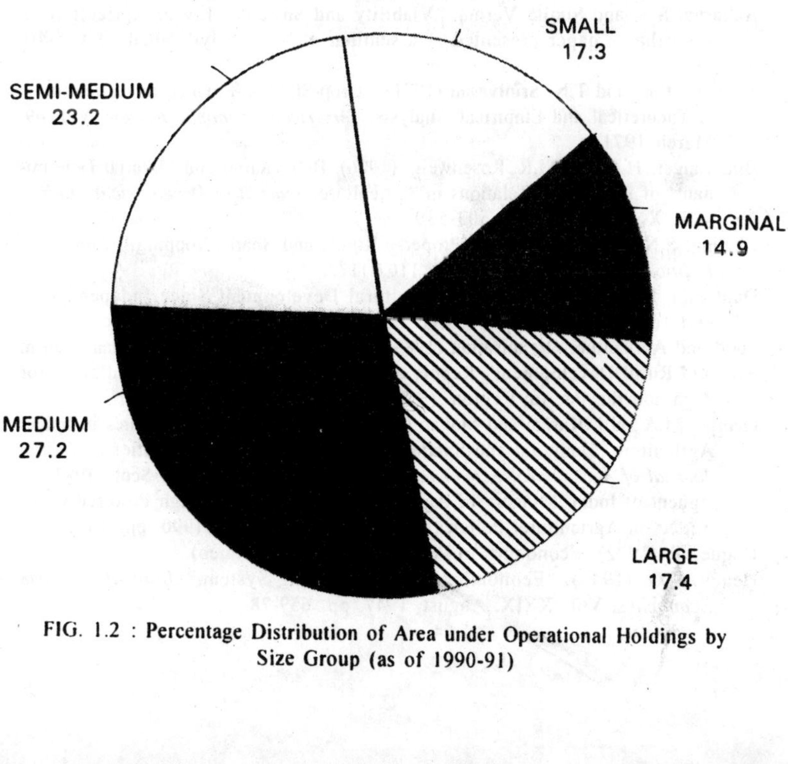

FIG. 1.2 : Percentage Distribution of Area under Operational Holdings by Size Group (as of 1990-91)

on the economics of various agricultural enterprises (other than crop-cultivation). Besides, these data do not adequately meet our objectives of inquiry into the questions concerning : (i) viability and sustainability of small and marginal farms; (ii) size-productivity relationship; (iii) role of various technological and institutional reforms in influencing both agricultural and non-agricultural development. Therefore, additional data were collected from other secondary sources including reports of ICAR, National Demonstrations and selected state government sources. Moreover, a comprehensive sample survey was conducted in the states of Andhra Pradesh, Bihar, Rajasthan and Haryana covering two representative districts from each state. Further, duly structured questionnaire schedules were pre-tested and canvassed in selected relatively developed as well as relatively backward villages of each selected district, covering a total of 900 farm households. Both tabular and regression analysis techniques were used to arrive at results and conclusions.

REFERENCES

Acharya, S.S. and Sunita Verma. "Viability and Sustainability of Agriculture in Rajasthan," paper presented at a seminar at NIRD, Hyderabad, Oct. 8-10, 1990.

Bardhan, P.K. and T.N. Srinivasan (1971). "Crop-Sharing Tenancy in Agriculture : A Theoretical and Empirical Analysis," *American Economic Review*, Vol. 69. March 1971.

Binswanger, H.P. and M.R. Rosenweig, (1986). Behavioural and Material Determinants of Production Relations in Agriculture, *Journal of Development Studies*, Vol. XXII, No. 3, pp. 503-539.

Cheung, S.N.S. (1968). "Private Property Rights and Share Cropping," *Journal of Political Economy*, Vol. 76, pp. 1107-1122.

Dantwala, M.L. (1991). Indian Agricultural Development Since Independence, Oxford-IBH Publishing Co. Pvt. Ltd., (Edited), p. 70.

Food and Agriculture Organisation (1989). "World Conference on Agrarian Reform and Rural Development" Report of the Sixth Government Consultation for Asia and the Pacific Region, (RAPA, 1989/7), pp. 21-22.

George, M.A., N.J. Kurian and Mohan C., Chandra (1983). "Factor Shares in Indian Agriculture, Temporal and Spatial Variations and their Implications," *Indian Journal of Agricultural Economics*, Vol. XXXVIII, No. 3, July-Sept., 1983.

Government of India, Ministry of Agriculture; "Report of the High Powered Committee on Agricultural Policies and Programmes," July, 1990, pp. 18-27.

Haque, T. (1992), Economics of Agriculture, NIRD (Mimeo).

Heady, E.O. (1947). "Economics of Farm Leasing System," *Journal of Farm Economics*, Vol. XXIX, August, 1947, pp. 659-78.

Hopper, W. David (1965). "Allocation Efficiency in Traditional Indian Agriculture." *Journal of Farm Economics*, Vol. 47, No. 3, August, 1965, pp. 611-624.

IP, P.C. and C.W. Stahl (1978). "Systems of Land Tenure, Allocative Efficiency and Economic Development, *American Journal of Agricultural Economics*, Vol. 60, No. 1, 1978, pp. 19-28.

Kaneda, Hiromitsu, "Structural Change and Policy Response in Japanese Agriculture after the Land Reform," *Economic Development and Cultural Change*, Vol. 28(3), 1980.

Leibenstein, H. (1966). "Allocative Efficiency Vs. X-Efficiency," *American Economic Review*, Vol. 61, pp. 392-415.

Lewis, W.A. (1955), "The Theory of Economic Growth," London, George Allen & Unwin, 1955.

Maji, C.C. and T. Haque (1990), "Economics of Marginal and Small Farms in West Bengal," Paper presented at the National Seminar on Economics of Agriculture with special reference to viability and sustainability of small and marginal farmers held at NIRD, Hyderabad, Oct. 8-10, 1990.

Parthasarathy, G. (1987). "Changes in the Incidence of Rural Poverty and Recent Trends in Some Aspects of Agrarian Economy," *Indian Journal of Agricultural Economics*, Vol. XXXXII, Jan.-March, 1987, pp. 1-21.

Raj Krishna (1964). "Some Production Functions for the Punjab," *Indian Journal of Agricultural Economics*, Vol. XIX, No. 3, July-Dec., 1964, pp. 87-97.

Rao, C.H.H. (1975). "Technological Change and Distribution of Gains in Indian Agriculture," New Delhi, 1975.

Rao, C.H.H. (1971). "Uncertainty, Entrepreneurship and Share Cropping in India," *Journal of Political Economy*, Vol. 79, May-June, 1971, pp. 578-95.

Reid, Jur., Jr. "Share Cropping in History and Theory," *Agricultural History*, Vol. 49, 1975, pp. 426-440.

Sen, A.K. (1962). "An Aspect of Indian Agriculture," *The Economic Weekly*, Vol. XIV, Nos. 4, 5&6, Nov.-Feb., 1962.

Swaminathan, M.S. (1983). "Science and the Conquest of Hunger," Concept Publishing Co., New Delhi.

2

PROFITABILITY OF AGRICULTURAL ENTERPRISES

Profitability is one of the important yardsticks for measuring farm efficiency. Any individual farmers' decision to allocate his limited resources among various enterprises and to invest in modern inputs does necessarily depend on whether such investment would generate sufficient profit. Besides, viability and sustainability of any farm enterprise depends to a large extent on its capacity to earn profit, although in a traditional agricultural economy of India, various other socio-economic considerations also enter into the decision-making process. Since farming is generally beset with lot of risks and uncertainties due to fluctuations in rainfall, prices and also in the availability of critical inputs in time like technology, credit etc., accrual of high rate of return in normal year/season becomes all the more important. However, the rate of profit of a farm enterprise is dependent not only on its capacity to obtain physical yields, but also on the behaviour of input-output prices which in turn depend on market and policy variables. Moreover, in a developing, albeit semi-controlled economy like ours, farm prices are often kept artificially low because of various socio-economic considerations, even though the policy of agricultural price protection is said to have operated to the disadvantage of surplus farm producers.

In recent years, however, there has emerged a strong farmer's lobby in the country which is demanding higher prices of agricultural output and subsidized prices of agricultural inputs, notwithstanding a global demand for reduction in subsidies. Some economists have indicated declining trends in profitability in agriculture in various regions of the country. If this proves to be true, then it is a matter of serious concern because the rate of private investment and consequently agricultural growth is dependent on the rate of profit. Keeping in view some of these facts, Chapter 2 is devoted to the economic analysis of various agricultural enterprises including crop production, animal husbandry, horticulture, sericulture, acquaculture etc. Also, the possibilities of raising farm income through changes in technology, diversification and appropriate land use planning have been examined.

Economics of Crop Production

In order to examine the general profitability of crop enterprises, time series secondary data, as collected from the Directorate of Economics and Statistics, Ministry of Agriculture, Government of India (1991) were analysed. This was supplemented through collection and analysis of farm level data from the four states of Bihar, Rajasthan, Andhra Pradesh and Haryana, covering eight districts and 15 villages including both developed and underdeveloped ones. Besides, the possibility of raising farm income through technological change was examined by analysing the data of ICAR national demonstrations and operational research programmes.

Crop production accounts for nearly 28 per cent of India's net national product, engaging about 78 per cent of the total rural workforce in the country. However, economics of crop production varies widely from region to region. Although non-availability of sufficient data for various agro-climatic zones prevents us from analysing the economics of crop production in various regions, the available statewise published data are quite revealing. Besides, results of our farm level survey enable us to analyse various economic aspects of crop production in the selected progressive as well as backward regions in a comparative framework.

It may be seen from Tables 2.1 and 2.2 that there are large scale inter-state variations in the cropping patterns as well as crop yields. Rice is the principal crop of a number of states including Andhra Pradesh, Assam, Bihar, Madhya Pradesh, Orissa, Tamil Nadu and West Bengal. Besides, rice is the second most important crop in states like Haryana, Punjab, Jammu & Kashmir and Kerala. It is an important crop in all other states too, excepting Gujarat. But the yields of rice ranged from 9 quintals per hectare in the states of Bihar and Rajasthan to 35 quintals in Punjab. The states of Assam, Bihar, Himachal Pradesh, Madhya Pradesh. Orissa and Rajasthan had relatively very low yields of rice than the national average of 19 quintals per hectare. Wheat is predominantly grown in Haryana, Himachal Pradesh, Punjab and Uttar Pradesh. It is also grown as the second most important crop in Bihar and Madhya Pradesh. But in most of the states, the average yields of wheat are very low as compared to the national average yield of 24 quintals per hectare. Also, small farmers prefer to grow cereals and pulses for the sake of food security and other conveniences. Kerala is the only state where plantation crops dominate the agricultural scene and sustain many small and marginal farmers. However, Table 2.3 shows that some major states like Uttar Pradesh, West Bengal, Haryana, Madhya Pradesh and Rajasthan have witnessed impressive growth rates of yields of foodcrops in recent years.

It may be seen from Table 2.4 that of the cereal crops, rice and wheat generate relatively higher net returns than maize, bajra and jowar. But the net returns from wheat cultivation are comparatively lower than that of rice in the states of Haryana, Madhya Pradesh and Punjab. Cash crops like sugarcane, cotton, etc. generate higher returns than the cereal crops. But relatively higher net returns from rice and wheat in Haryana and Punjab also demonstrate that given the access to technology and other infrastructural and institutional facilities, even cereal crops have enough potentials which need to be fully explored and utilised. In fact certain regions seem to have distinct economic advantages in terms of net returns from crops like Haryana and Punjab for rice and wheat, Maharashtra for jowar, Bihar and Rajasthan for maize, Gujarat for bajra, Rajasthan for gram, Uttar Pradesh for potato, Punjab for cotton, and Karnataka, Andhra Pradesh and Maharashtra for sugarcane. Also the net profitability in crop cultivation was found to be very low for all the crops when measured at cost C_2 which includes operational expenses plus rental value of owned land, rent paid for leased in land, interest on value of owned fixed capital assets (excluding land) and imputed value of family labour.

Besides, there is lot of instability in the rate of returns. Table 2.5 shows that assuming 1971-72 as the base, the index of factor—productivity of rice declined in all the states excepting Haryana. Table 2.6 further shows that over time, the index of factor productivity of wheat declined in all the states excepting Rajasthan, which confirms that cost of cultivation is rising at a faster rate than that of returns from wheat cultivation.

TABLE 2.1 : Statewise Cropping Patterns as of 1990-91

State	Cropping Pattern (area in 000'ha)									
Andhra Pradesh	Rice	G. Nut	Pulses	Jawar	Cotton	Maize	Bajra	Chillies	S. Cane	Tobacco
Area	4006	2389	1637	1245	657	318	235	203	179	167
Assam	Rice	Mustard	Pulses	Jute	Wheat	Arcanut	Potato	S. Cane	Maize	
Area	2491	295	111	96	84	63	59	36	20	
Bihar	Rice	Wheat	Pulses	Maize	Potato	S. Cane	Jute	Mustard	Tobacco	
Area	5393	1965	1187	663	161	149	137	109	15	
Gujarat	G. Nut	Bajra	Pulses	Cotton	Jowar	Mustard	Maize	Castor seed	Sesamum	
Area	1702	1152	932	921	853	417	369	345	239	
Haryana	Wheat	Pulses	Rice	Bajra	Cotton	Rape seed & Mustard		S. Cane	Jowar	Potato
Area	1851	732	667	608	489	471		147	129	10
Himachal Pradesh	Wheat	Maize	Rice	Pulses	Barley	Ragi & S. Millets		Potato		
Area	376	319	85	43	29	22		16		
Jammu & Kashmir	Maize	Rice	Wheat	Mustard	Pulses					
Area	296	273	226	59	38					
Karnataka	Jowar	Pulses	G. Nut	Rice	Sun Flower	Cotton	Bajra	S. Cane	Maize	Coconut
Area	2155	1483	1186	1167	1057	596	425	257	246	233

(*Contd.*)

TABLE 2.1 : (*Contd.*)

State	Cropping Pattern (area in 000'ha)									
Kerala	Coconut	Rice	Pepper	Tapioca	Arcanut	Cardamom	Banana	Pulses	G. Nut	Ginger
Area	864	560	169	147	62	62	61	25	14	14
Madhya Pradesh	Rice	Pulses	Wheat	Jowar	Maize	Cotton	Mustard	G. Nut	Bajra	
Area	5062	4798	3739	1732	889	587	506	360	174	
Maharashtra	Jowar	Pulses	Cotton	Bajra	Rice	G. Nut	Wheat	Safflower	Sunflower	S. Cane
Area	6331	3258	2730	1927	1582	881	873	628	497	444
Orissa	Rice	Pulses	G. Nut	Maize	Mustard	Chilies	Jute & Mesta	S. Cane	Wheat	Coconut
Area	4404	2204	398	167	162	90	70	49	34	33
Punjab	Wheat	Rice	Cotton	Maize	Pulses	S. Cane	Mustard	Potato	Bajra	G. Nut
Area	3272	2024	701	188	148	101	73	19	11	10
Rajasthan	Bajra	Pulses	Guav Seed	Mustard	Wheat	Maize	Jowar	Sesamum	Cotton	Barley	G. Nut
Area	4846	3686	2099	2042	1814	982	929	583	455	234	231
Tamil Nadu	Rice	G. Nut	Pulses	Jowar	Bajra	S. Cane	Coconut	Cotton	Ragi & S. Millets	Barley
Area	1873	940	861	545	264	239	226	217	341	140
Uttar Pradesh	Wheat	Rice	Pulses	S. Cane	Maize	Mustard	Bajra	Jowar	Ragi & S. Millets	Barley
Area	8490	5585	3010	1586	1133	1124	766	527	507	431
West Bengal	Rice	Jute	Mustard	Pulses	Wheat	Potato	Sesamum	Maize		
Area	5813	500	378	357	269	194	99	65		

TABLE 2.2 : Yields of Major Crops by State (As of 1993-94)

(*Yields Tonnes/ha*)

State	Rice	Wheat	Jowar	Bajra	Maize	Total Food-grains	S. Cane	Cotton	Rape Seed & Mustard
(1)	(2)	(3)	(4)	(5)	(6)	(7)	(8)	(9)	(10)
All India	1.9	2.4	0.9	0.5	1.6	1.5	67.1	0.2	0.9
Andhra Pradesh	2.7	0.8	0.8	0.8	2.6	1.8	77.3	0.3	0.3
Assam	1.3	1.5	—	—	0.6	1.3	38.4	—	0.5
Bihar	0.9	2.1	0.5	0.4	2.0	1.5	36.7	—	0.8
Gujarat	1.4	1.9	0.7	0.6	1.0	0.9	79.7	0.2	0.9
Haryana	2.8	3.6	0.3	0.7	1.6	2.6	57.6	0.3	1.4
Himachal Pradesh	1.2	1.1	—	—	2.1	1.4	13.4	—	—
Jammu & Kashmir	1.9	1.4	—	0.5	1.7	1.6	29.8	—	0.6
Karnataka	2.3	0.7	0.9	0.6	2.8	1.2	85.4	0.2	—
Kerala	1.9	—	—	—	—	1.8	—	—	—
Madhya Pradesh	1.2	1.6	1.0	0.8	1.5	1.1	35.4	0.1	0.9
Maharashtra	1.6	1.4	1.0	0.7	1.7	0.9	81.0	0.2	—
Orissa	1.5	1.7	—	—	1.0	1.3	58.3	—	0.5
Punjab	3.5	4.0	—	1.1	1.9	3.7	61.2	0.4	1.1
Rajasthan	0.9	1.7	0.2	0.2	1.0	0.7	49.5	0.3	0.7
Tamilnadu	2.8	—	1.1	1.4	1.5	1.9	105.4	0.3	—
Uttar Pradesh	1.9	2.3	0.9	1.1	1.2	1.8	59.1	0.2	1.0
West Bengal	2.0	2.1	—	—	3.2	2.0	57.8	—	0.8

TABLE 2.3 : Productivity Growth in Selected Crops (1981-82 : 1991-92)
Annual Growth Rates of Yields (Per cent)

State	Total Food-grains	Rice	Wheat	Jowar	Maize	Bajra	Cotton	Sugar Cane	Rape Seed & Mustard
(1)	(2)	(3)	(4)	(5)	(6)	(7)	(8)	(9)	(10)
All India	2.5	3.5	3.6	1.2	3.2	2.9	3.8	1.7	5.7
Andhra Pradesh	3.3	2.1	4.6	1.1	1.4	0.2	1.5	–0.9	5.0
Assam	2.2	2.2	0.7	—	0.4	—	—	0.7	1.2
Bihar	3.6	2.9	3.4	–1.8	6.1	–0.4	—	4.7	5.7
Gujarat	–0.2	1.1	0.4	–3.1	2.7	–1.3	1.2	4.1	9.4
Haryana	5.2	1.9	4.0	2.0	4.6	3.8	2.3	–0.5	2.6
Himachal Pradesh	2.8	1.6	3.6	—	1.9	—	—	–3.5	—
Jammu & Kashmir	0.2	0.4	1.6	—	–0.6	–2.1	—	2.9	–7.3
Karnataka	–0.3	0.4	–0.7	–1.8	0.6	2.8	8.6	0.3	—
Kerala	1.5	1.7	—	—	—	—	—	—	—
Madhya Pradesh	4.2	4.4	4.8	3.5	4.5	6.4	2.6	3.2	6.0
Maharashtra	1.0	–0.2	1.8	1.5	–3.8	2.5	1.6	–0.9	—
Orissa	3.7	44.	–0.6	—	3.2	—	—	1.4	2.5
Punjab	3.0	1.9	2.8	—	0.9	–0.5	5.6	1.3	4.4
Rajasthan	4.2	3.2	4.6	2.3	4.2	8.9	5.8	2.4	3.8
Tamilnadu	2.0	3.7	—	2.5	0.2	2.1	3.5	–0.1	—
Uttar Pradesh	4.7	7.2	3.6	4.8	4.8	6.0	2.5	2.6	6.7
West Bengal	4.4	4.4	1.3	—	6.6	—	—	1.9	4.4

TABLE 2.4 : Per Hectare Net Returns from Principal Crops

Crop	State	Net Returns (Rs./Ha.)							
		At A_1				At C_2			
		1983-84	1986-87	1987-88	1990-91	1983-84	1986-87	1987-88	1990-91
(1)	(2)	(3)	(4)	(5)	(6)	(7)	(8)	(9)	(10)
Rice	Andhra Pradesh	2121	NR	3218	NR	122	NR	405	NR
	Assam	1935	1439	NR	NR	649	51	NR	NR
	Bihar	2744	NR	2748	NR	1087	NR	602	NR
	Haryana	2288	3697	4882	6843	352	1829	2400	3211
	Karnataka	4095	5390	NR	NR	2166	3216	NR	NR
	Madhya Pradesh	1658	1769	NR	NR	783	617	NR	NR
	Orissa	2432	1856	NR	NR	1020	457	NR	NR
	Punjab	3730	4895	5330	6323	1012	1494	1760	1268
Wheat	Haryana	1937	2731	NR	8938	302	961	NR	5590
	Madhya Pradesh	1506	1592	NR	2575	542	426	NR	786
	Punjab	2546	2801	NR	5440	556	409	NR	1439
	Rajasthan	2245	4643	NR	NR	825	2521	NR	NR
	Uttar Pradesh	1992	2936	NR	NR	471	1252	NR	NR
Jowar	Andhra Pradesh	NR	510	NR	NR	NR	–147	NR	NR
	Madhya Pradesh	822	928	NR	NR	263	171	NR	NR
	Maharashtra	1038	1420	NR	NR	342	573	NR	NR

(*Contd.*)

TABLE 2.4 : (*Contd.*)

(1)	(2)	(3)	(4)	(5)	(6)	(7)	(8)	(9)	(10)
Maize	Madhya Pradesh	876	799	NR	NR	238	–15	NR	NR
Bajra	Gujarat	1384	2378	NR	NR	465	982	NR	NR
	Haryana	685	1432	NR	NR	–182	96	NR	NR
Cotton	Karnataka	7205	1442	NR	NR	1261	202	NR	NR
	Punjab	1178	4124	NR	NR	543	1231	NR	NR
Sugar Cane	Andhra Pradesh	7674	12158	NR	NR	2518	5792	NR	NR
	Haryana	4243	8909	NR	NR	1534	6045	NR	NR
	Karnataka	15228	19568	NR	NR	10705	14008	NR	NR
	Uttar Pradesh	NR	7856	NR	NR	NR	4665	NR	NR
Potato	Uttar Pradesh	6569	9949	NR	NR	3693	6954	NR	NR

Table 2.5 to 2.18 shows that there has been a general declining trend in factor productivity in most of the regions for all the crops, excepting potato and sugarcane in general and wheat in Haryana and Punjab. This could be largely attributed to rising factor prices and not so much due to increased use of inputs. Tables 2.7 and 2.8 indicate how the rising cost of fertilizer and human labour have affected factor productivity overtime. The only way to meet the situation could be through further technological improvement, particularly in the direction of cost effective technology. Further, it could be seen from Fig. 2.1 and 2.2 that the wages rates of agricultural labour have been rising at faster rates than those of employment.

Changes in the Structure of Cost of Cultivation

Table 2.19 shows how the structure of cost of cultivation remains more or less traditional in the sense that wages of human labour constituted the highest percentage share in the total operational expenses in the case of all cereal crops excepting wheat in all the major states for which data were available. Figs. 2.3 to 2.6 indicate the variations in the percentage shares of human labour and fertilizers in the total operational expenses of rice and wheat cultivation (for details see Haque, 1992).

TABLE 2.5 : Changes in Factor—Productivity of Rice

State	Year	Index of input at Cost A_1	Index of input at Cost C_2	Index of output	Index of factor productivity at Cost A_1	Index of factor productivity at Cost C_2
(1)	(2)	(3)	(4)	(5)	(6)	(7)
Andhra Pradesh	1971-72	100	100	100	100	100
	1972-73	92	92	93	101	101
	1973-74	113	116	131	116	113
	1978-79	241	209	164	68	78
	1983-84	400	358	310	78	87
	1987-88	535	489	436	81	89
	1988-89	602	577	544	90	94
Assam	1971-72	100	100	100	100	100
	1972-73	109	83	108	99	130
	1973-74	120	92	121	101	132
	1978-79	198	142	169	85	119
	1983-84	316	253	293	93	116
	1986-87	399	298	270	68	91

(*Contd.*)

TABLE 2.5 : (*Contd.*)

(1)	(2)	(3)	(4)	(5)	(6)	(7)
Bihar	1972-73	100	100	100	100	100
	1973-74	116	118	122	105	103
	1983-84	206	238	226	110	95
	1987-88	298	323	256	86	79
Haryana	1984-85	100	100	100	100	100
	1985-86	107	107	126	118	118
	1986-87	103	103	128	124	124
	1987-88	103	105	151	147	144
	1988-89	117	112	173	148	154
	1990-91	164	173	214	130	124
Karnataka	1973-74	100	100	100	100	100
	1983-84	220	201	193	88	96
	1984-85	260	237	221	85	93
	1985-86	220	201	215	98	107
	1986-87	249	227	217	88	98
	1989-90	375	302	197	53	65
Madhya Pradesh	1972-73	100	100	100	100	100
	1973-74	126	129	145	115	112
	1983-84	266	240	268	101	112
	1984-85	283	264	210	74	80
	1985-86	413	361	379	92	105
	1986-87	420	352	333	79	95
	1987-88	518	528	472	91	89
	1990-91	680	612	544	80	89
Orissa	1971-72	100	100	100	100	100
	1972-73	105	105	107	102	102
	1973-74	87	95	101	116	106
	1978-79	171	156	136	80	87
	1983-84	339	333	346	102	104
	1984-85	387	336	312	81	93
	1986-87	434	378	329	76	87
	1990-91	246	295	266	108	90
Punjab	1978-79	100	100	100	100	100
	1983-84	174	190	176	101	93
	1984-85	189	205	176	93	86
	1985-86	176	194	183	104	94
	1986-87	184	216	208	113	96
	1987-88	198	230	225	114	98
	1988-89	195	225	225	115	100
West Bengal	1971-72	100	100	100	100	100
	1972-73	109	100	107	98	107
	1973-74	105	107	120	114	112
	1978-79	272	205	162	60	79
	1983-84	390	330	310	79	94
	1984-85	377	338	321	85	95
	1989-90	604	520	449	74	86

TABLE 2.6 : Changes in Factor—Prodctivity of Wheat

State	Year	Index of input at Cost A_1	Index of input at Cost C_2	Index of output	Index of factor productivity at Cost A_1	Index of factor productivity at Cost C_2
(1)	(2)	(3)	(4)	(5)	(6)	(7)
Haryana	1971-72	100	100	100	100	100
	1973-74	126	124	130	103	105
	1978-79	281	243	161	57	66
	1983-84	361	320	136	38	43
	1984-85	370	337	268	72	80
	1985-86	405	362	322	80	89
	1986-87	412	358	297	72	83
	1987-88	410	389	344	84	88
	1988-89	457	392	389	85	99
	1991-92	637	602	714	112	119
Madhya Pradesh	1971-72	100	100	100	100	100
	1972-73	126	124	128	102	103
	1973-74	153	158	167	109	106
	1983-84	346	328	311	90	95
	1984-85	354	366	339	96	93
	1985-86	356	357	339	95	95
	1986-87	395	384	340	86	86
	1988-89	610	591	539	88	91
	1991-92	995	973	924	93	95
Punjab	1971-72	100	100	100	100	100
	1972-73	91	93	83	91	89
	1973-74	115	115	129	112	112
	1978-79	183	172	153	84	89
	1983-84	257	252	226	88	90
	1984-85	286	291	276	97	95
	1985-86	298	305	305	102	100
	1986-87	304	300	258	85	86
	1987-88	333	336	331	99	99
	1988-89	355	378	373	105	99
	1990-91	418	452	476	114	105
Rajasthan	1983-84	100	100	100	100	100
	1984-85	96	114	118	123	104
	1985-86	117	131	164	140	125
	1986-87	112	129	166	148	129
	1989-90	163	174	202	124	116
Uttar Pradesh	1971-72	100	100	100	100	100
	1972-73	106	111	95	90	86
	1973-74	122	125	134	110	107
	1978-79	186	181	145	78	80
	1983-84	277	261	213	77	99
	1985-86	317	284	257	81	90
	1986-87	325	299	281	86	94

TABLE 2.7 : Changes in the Index of Input Prices and their Uses in Rice Cultivation
(Base : 1981-82 = 100)

State	Wage Rate of Labour			Use of Human Labour			Price of Fertiliser			Use of Fertiliser		
	1986-87	1987-88	1988-89	1986-87	1987-88	1988-89	1986-87	1987-88	1988-89	1986-87	1987-88	1988-89
Andhra Pradesh	—	190	—	—	113	—	—	96	—	—	132	—
Assam	190	—	—	95	—	—	95	—	—	1454	—	—
Bihar	—	176	—	—	118	—	—	92	—	—	222	—
Haryana*	115	134	141	104	112	108	104	92	93	92	108	102
Karnataka	144	—	—	79	—	—	94	—	—	105	—	—
Madhya Pradesh	160	—	—	93	—	—	99	—	—	142	—	—
Orissa	170	—	—	100	—	—	85	—	—	402	—	—
Punjab	132	128	166	98	94	82	107	90	96	97	43	104

* for Haryana, the base year refer to 1984-85.

TABLE 2.8 : Changes in the Index of Input Prices and their Uses in Wheat Cultivation
Base Year 1974-75 = 100 (Wheat)

State	Wage of Human Labour			Use of Human Labour			Price of Fertiliser			Use of Fertiliser		
	1983-84	1986-87	1988-89	1983-84	1986-87	1988-89	1983-84	1986-87	1988-89	1983-84	1986-87	1988-89
Haryana	245	257	297	75	68	76	98	104	108	421	529	576
Madhya Pradesh	225	275	288	64	66	76	77	83	—	127	155	259
Punjab	171	223	277	73	75	69	92	94	95	201	227	230
Rajasthan	183	233	—	78	67	—	119	126	—	148	127	—
Uttar Pradesh	200	280	—	81	74	—	120	123	—	166	189	—

TABLE 2.9 : Changes in Factor—Productivity of Jowar

State	Year	Index of input at Cost A_1	Index of input at Cost C_2	Index of output	Index of factor productivity at Cost A_1	Index of factor productivity at Cost C_2
(1)	(2)	(3)	(4)	(5)	(6)	(7)
Andhra Pradesh	1973-74	100	100	100	100	100
	1984-85	344	310	220	64	71
	1986-87	338	314	240	71	76
Karnataka	1971-72	100	100	100	100	100
	1972-73	137	126	71	52	56
	1973-74	133	132	87	65	66
	1978-79	197	161	73	37	45
	1983-84	432	343	180	42	52
	1987-88	545	413	291	53	70
Madhya Pradesh	1978-79	100	100	100	100	100
	1983-84	183	161	162	89	101
	1984-85	193	181	155	80	86
	1985-86	193	191	183	95	96
	1986-87	215	203	186	87	92
	1987-88	251	261	235	94	90
	1990-91	466	438	349	75	80
Maharashtra	1971-72	100	100	100	100	100
	1972-73	114	85	86	75	101
	1973-74	181	138	144	80	104
	1978-79	231	152	148	64	97
	1983-84	491	337	308	73	91
	1986-87	659	434	418	63	96
Rajasthan	1971-72	100	100	100	100	100
	1972-73	137	126	112	82	88
	1973-74	133	132	138	104	105
	1978-79	170	148	115	68	78
	1983-84	432	343	285	66	83

TABLE 2.10 : Changes in Factor—Productivity of Bajra
(Base 1973-74 = 100)

Item	Year	State		
		Gujarat	Haryana	Rajasthan
Index of Cost A_1	1973-74	100	100	100
	1983-84	343	152	261
	1986-87	450	259	NR
Index of Cost C_2	1973-74	100	100	100
	1983-84	291	157	264
	1986-87	405	251	NR
Index of Output	1973-74	100	100	100
	1983-84	232	109	204
	1986-87	350	211	NR
Index of Factor Productivity at A_1	1973-74	100	100	100
	1983-84	68	72	78
	1986-87	78	81	NR
Index of Factor Productivity at C_2	1983-84	78	69	77
	1986-87	86	84	NR

TABLE 2.11 : Changes in Factor—Productivity of Maize

State	Year	Index of input at Cost A_1	Index of input at Cost C_2	Index of output	Index of factor productivity at Cost A_1	Index of factor productivity at Cost C_2
(1)	(2)	(3)	(4)	(5)	(6)	(7)
Madhya Pradesh	1983-84	100	100	100	100	100
	1985-86	135	141	119	88	84
	1986-87	120	124	103	86	83
Rajasthan	1972-73	100	100	100	100	100
	1973-74	83	83	100	120	120
	1983-84	240	259	293	120	113
	1984-85	297	305	302	102	99
	1985-86	305	336	386	127	115

TABLE 2.12 : Changes in Factor—Productivity of Gram

State	Year	Index of input at Cost A_1	Index of input at Cost C_2	Index of output	Index of factor productivity at Cost A_1	Index of factor productivity at Cost C_2
(1)	(2)	(3)	(4)	(5)	(6)	(7)
Haryana	1978-79	100	100	100	100	100
	1983-84	120	120	95	79	79
	1984-85	138	141	135	98	96
	1985-86	146	164	190	130	79
	1986-87	153	162	156	101	96
	1987-88	183	200	165	90	83
Madhya Pradesh	1978-79	100	100	100	100	100
	1983-84	154	154	136	88	88
	1984-85	208	206	188	90	91
	1985-86	211	199	188	89	94
	1986-87	211	187	157	74	84
	1987-88	261	244	220	84	90
Rajasthan	1983-84	100	100	100	100	100
	1984-85	117	118	132	113	113
	1985-86	126	155	199	158	128
	1986-87	119	129	160	134	124
Uttar Pradesh	1983-84	100	100	100	100	100
	1984-85	112	116	115	103	99
	1985-86	120	115	124	103	108
	1986-87	116	114	98	84	86

TABLE 2.13 : Changes in Factor—Productivity of Potato

State	Year	Index of input at Cost A_1	Index of input at Cost C_2	Index of output	Index of factor productivity at Cost A_1	Index of factor productivity at Cost C_2
(1)	(2)	(3)	(4)	(5)	(6)	(7)
Bihar	1981-82	100	100	100	100	100
	1982-83	112	100	97	87	97
	1983-84	101	93	88	87	95

(*Contd.*)

TABLE 2.13 (*Contd.*)

(1)	(2)	(3)	(4)	(5)	(6)	(7)
Himachal Pradesh	1975-76	100	100	100	100	100
	1981-82	121	139	154	127	111
	1982-83	125	126	128	102	102
	1983-84	149	151	157	105	104
Uttar Pradesh	1981-82	100	100	100	100	100
	1982-83	122	117	159	130	136
	1983-84	140	133	186	133	140
	1984-85	113	110	91	81	83
	1985-86	110	115	217	197	189

TABLE 2.14 : Changes in Factor—Productivity of Groundnut

State	Year	Index of input at Cost A_1	Index of input at Cost C_2	Index of output	Index of factor productivity at Cost A_1	Index of factor productivity at Cost C_2
(1)	(2)	(3)	(4)	(5)	(6)	(7)
Gujarat	1973-74	100	100	100	100	100
	1978-79	139	129	112	81	87
	1983-84	287	266	250	87	94
	1986-87	342	321	352	103	110
Karnataka	1983-84	100	100	100	100	100
	1985-86	114	110	98	86	89
	1986-87	137	125	116	85	93
Orissa	1984-85	100	100	100	100	100
	1985-86	110	100	86	78	86
	1986-87	120	109	95	79	87
Tamil Nadu	1973-74	100	100	100	100	100
	1978-79	117	111	87	74	78
	1983-84	201	198	184	92	93

TABLE 2.15 : Changes in Factor—Productivity of Sugarcane

State	Year	Index of input at Cost A_1	Index of input at Cost C_2	Index of output	Index of factor productivity at Cost A_1	Index of factor productivity at Cost C_2
Andhra Pradesh	1978-79	100	100	100	100	100
	1983-84	169	187	208	123	111
	1984-85	159	189	230	145	122
	1985-86	173	213	302	175	142
	1986-87	192	219	287	149	131
	1987-88	213	222	240	113	108
Haryana	1983-84	100	100	100	100	100
	1984-85	112	103	120	107	117
	1985-86	121	113	163	135	144
	1986-87	144	122	189	131	155
Karnataka	1983-84	100	100	100	100	100
	1984-85	196	150	117	60	78
	1985-86	128	124	127	99	102
	1986-87	132	127	129	98	102
Maharashtra	1973-74	100	100	100	100	100
	1983-84	244	220	162	66	74
	1986-87	245	234	181	74	77
	1987-88	250	234	192	77	82
Tamil Nadu	1983-84	100	100	100	100	100
	1987-88	125	123	111	89	90
Uttar Pradesh	1973-74	100	100	100	100	100
	1984-85	157	163	168	107	107
	1985-86	146	155	212	145	145
	1986-87	177	176	213	120	120

TABLE 2.16 : Changes in Factor—Productivity of Cotton

State	Year	Index of input at Cost A_1	Index of input at Cost C_2	Index of output	Index of factor productivity at Cost A_1	Index of factor productivity at Cost C_2
Gujarat	1974-75	100	100	100	100	100
	1983-84	324	333	374	115	112

(Contd.)

TABLE 2.16 : (*Contd.*)

(1)	(2)	(3)	(4)	(5)	(6)	(7)
Haryana	1974-75	100	100	100	100	100
	1987-88	171	211	255	149	121
Karnataka	1983-84	100	100	100	100	100
	1985-86	57	58	43	75	74
	1986-87	52	52	30	58	58
Madhya Pradesh	1983-84	100	100	100	100	100
	1984-85	128	173	284	222	164
	1985-86	140	163	232	166	142
	1986-87	176	184	225	128	122
Maharashtra	1978-79	100	100	100	100	100
	1983-84	187	210	217	116	103
Punjab	1972-73	100	100	100	100	100
	1973-74	123	121	130	106	107
	1983-84	264	224	150	57	67
	1984-85	320	295	362	113	123
	1985-86	312	306	347	111	113
	1986-87	372	346	339	91	98
	1989-90	525	534	606	115	113

TABLE 2.17 : Changes in Factor—Productivity of Jute

State	Year	Index of input at Cost A_1	Index of input at Cost C_2	Index of output	Index of factor productivity at Cost A_1	Index of factor productivity at Cost C_2
(1)	(2)	(3)	(4)	(5)	(6)	(7)
Assam	1973-74	100	100	100	100	100
	1978-79	165	146	116	70	79
	1983-84	324	303	166	51	55
	1984-85	411	300	207	50	69
	1985-86	629	391	215	34	55
	1986-87	483	364	236	49	65
	1987-88	478	399	322	67	81
Bihar	1973-74	100	100	100	100	100
	1983-84	154	150	158	103	105
	1985-86	232	201	174	75	87
	1986-87	253	236	230	91	97

(*Contd.*)

TABLE 2.17 : (*Contd.*)

(1)	(2)	(3)	(4)	(5)	(6)	(7)
Orissa	1973-74	100	100	100	100	100
	1978-79	266	199	159	60	80
	1983-84	291	313	337	116	108
	1985-86	536	452	381	71	84
	1986-87	568	433	307	54	71
West Bengal	1973-74	100	100	100	100	100
	1978-79	202	189	149	74	79
	1983-84	226	238	275	122	116
	1984-85	242	296	459	190	155
	1985-86	313	270	241	77	77
	1986-87	309	276	254	82	92
	1987-88	361	326	278	77	85

TABLE 2.18 : Changes in Factor—Productivity of Rapeseed & Mustard

State	Year	Index of input at Cost A_1	Index of input at Cost C_2	Index of output	Index of factor productivity at Cost A_1	Index of factor productivity at Cost C_2
(1)	(2)	(3)	(4)	(5)	(6)	(7)
Assam	1978-79	100	100	100	100	100
	1983-84	207	226	220	106	97
	1984-85	222	199	142	64	71
	1985-86	245	238	163	67	68
	1986-87	251	256	193	77	75
Haryana	1984-85	100	100	100	100	100
	1985-86	123	112	92	75	82
	1986-87	140	122	117	84	96
	1987-88	127	131	177	139	135
Rajasthan	1983-84	100	100	100	100	100
	1984-85	123	124	121	98	98
	1985-86	77	99	117	152	118
	1986-87	112	132	190	170	144
Uttar Pradesh	1983-84	100	100	100	100	100
	1984-85	77	59	91	118	154
	1985-86	87	62	94	108	152
	1986-87	85	66	120	141	182

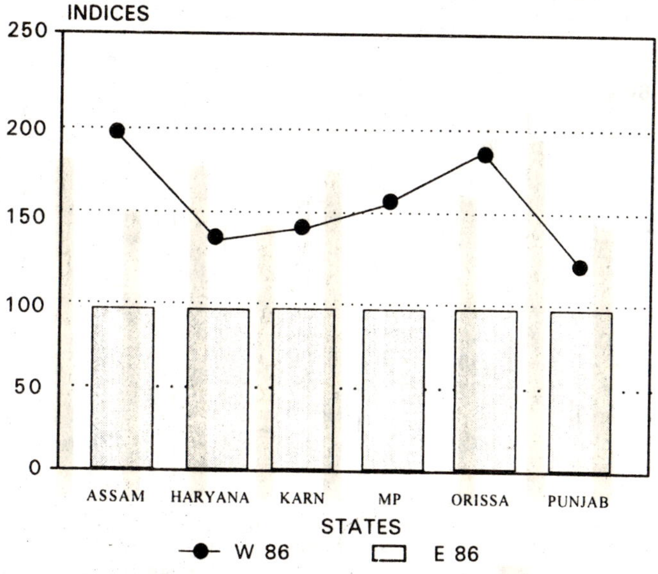

FIG. 2.1 : Index of Wage and Employment (Human Labour—Rice)

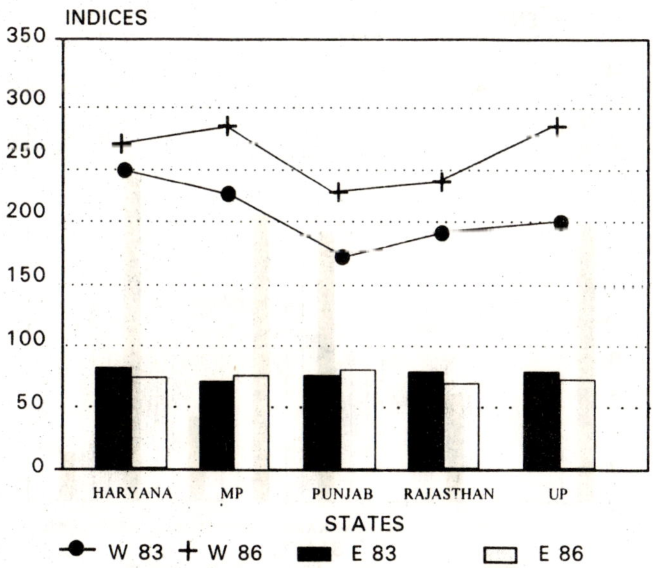

FIG. 2.2 : Index of Wage and Employment (Human Labour—Wheat)

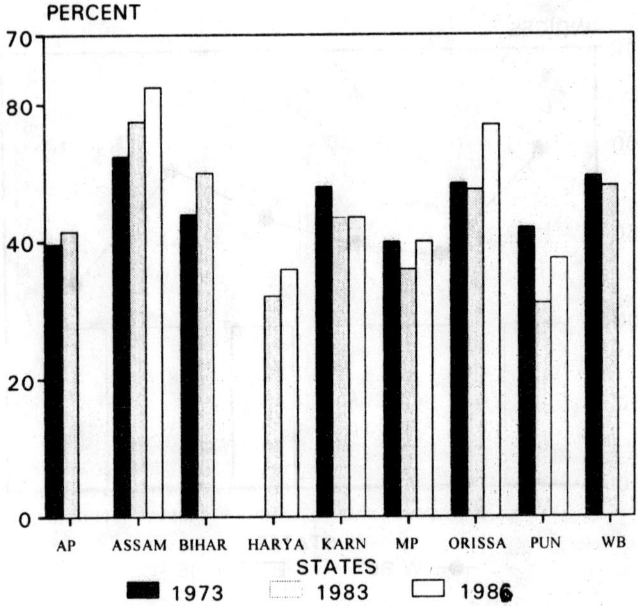

FIG. 2.3 : Human Labour in Total Operational Cost (Per cent Share—Rice)

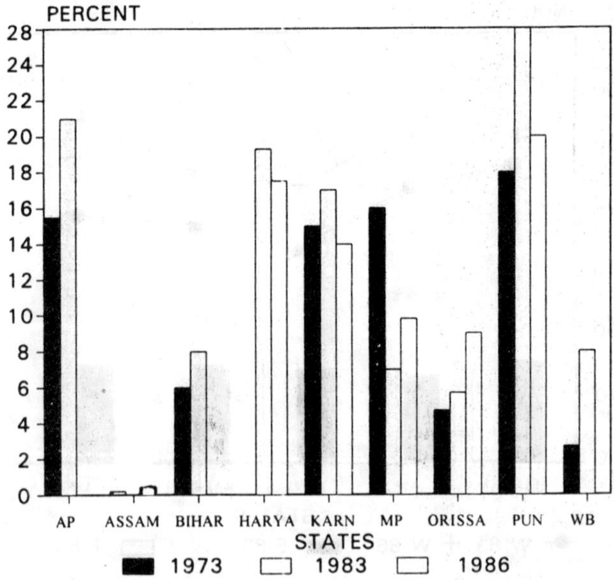

FIG. 2.4 : Fertiliser's Share in Operational Cost (Per cent Share—Rice)

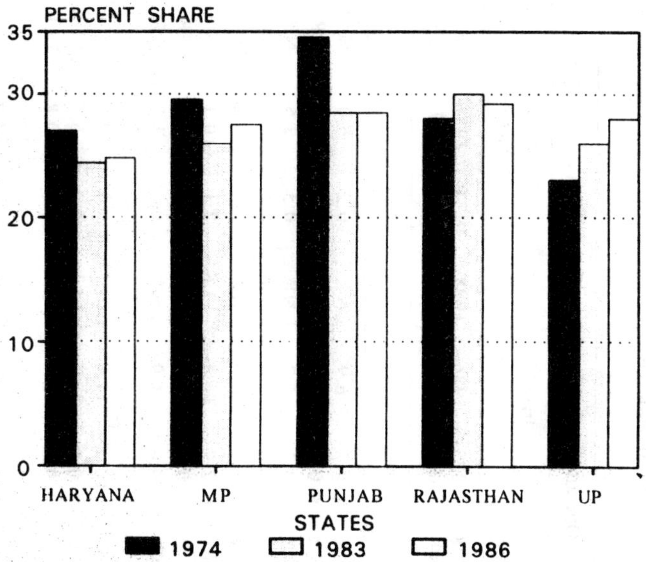

FIG. 2.5 : Human labour in Operational Cost (Per cent Share—Wheat)

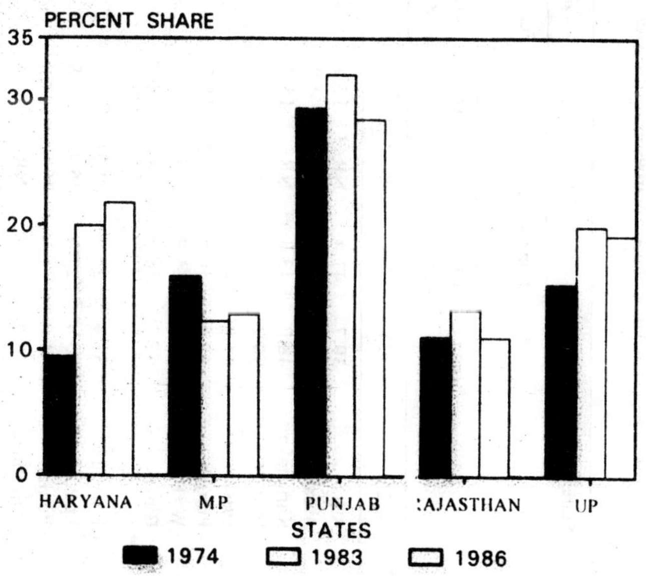

FIG. 2.6 : Fertiliser's Share in Operational Cost (Per cent Share—Wheat)

TABLE 2.19 : Variation in the Structure of Cost of Cultivation Over Time

Crop	State	% Share in Operational Cost											
		Human Labour			Fertiliser			Irrigation			Machine Labour		
		1973-74	1986-87	1988-89	1973-74	1986-87	1988-89	1973-74	1986-87	1988-89	1973-74	1986-87	1988-89
(1)	(2)	(3)	(4)	(5)	(6)	(7)	(8)	(9)	(10)	(11)	(12)	(13)	(14)
Paddy	A.P.	39.7	41.4	NR	17.4	18.1	NR	6.7	6.9	NR	10.2	6.0	NR
	Assam	51.2	62.6	NR	0.0	0.5	NR	0.0	0.0	NR	0.0	0.2	NR
	Karnataka	48.2	43.3	NR	15.2	13.9	NR	0.3	1.7	NR	0.5	2.6	NR
	M.P.	40.2	39.5	NR	16.2	9.9	NR	0.0	1.9	NR	0.0	0.0	NR
	Orissa	48.9	52.8	NR	4.3	9.3	NR	0.3	0.8	NR	0.0	0.0	NR
	Punjab	41.6	36.5	32.2*	18.3	20.9	20.8	18.7	17.4	17.6	2.1	10.1	14.6
	W. Bengal	49.1	55.9	NR	1.7	5.9	NR	0.2	1.1	NR	0.0	0.3	NR
	Bihar	43.7	51.3	NR	6.2	10.6	NR	0.0	1.9	NR	0.0	1.2	NR
Wheat	Haryana	31.9	25.2	21.7	9.3	23.8	23.7	12.1	11.8	13.1	7.3	19.4	20.4
	M.P.	35.4	27.3	26.1	10.8	12.8	14.6	0.1	12.5	14.1	0.7	6.5	10.5
	Punjab	31.0	26.9	26.6	24.0	28.5	25.6	9.1	5.8	5.7	11.3	20.8	22.9
	Rajasthan	28.0	28.7	NR	11.2	11.3	NR	16.6	16.2	NR	6.0	13.5	NR
	U.P.	29.0	28.0	NR	9.4	19.7	NR	14.1	9.5	NR	4.1	14.2	NR

Jowar	A.P.	45.5	43.1	NR	5.9	5.2	NR	0.9	1.9	NR	0.4	1.7	NR
	M.P.	53.4	43.8	NR	1.4	5.9	NR	0.0	0.2	NR	0.4	1.5	NR
	Maharashtra	45.8	41.7	NR	2.3	10.9	NR	1.1	3.4	NR	0.2	2.7	NR
Maize	Rajasthan	46.2	39.2	NR	2.9	8.9	NR	0.0	5.9	NR	0.0	1.7	NR
Bajra	Gujarat	38.7	40.0	NR	7.7	10.7	NR	8.3	16.6	NR	2.8	8.5	NR
	Haryana	57.6	47.0	NR	3.5	2.4	NR	3.7	5.4	NR	1.9	12.2	NR
Cotton	Gujarat	47.0	31.1	NR	9.2	4.0	NR	10.6	3.8	NR	1.5	2.9	NR
	Karnataka	30.0	34.3	NR	18.0	15.0	NR	0.2	2.8	NR	0.3	2.1	NR
	Punjab	50.3	56.7	NR	12.9	8.7	NR	4.9	5.7	NR	5.2	6.1	NR
Groundnut	Gujarat	29.3	24.6	NR	7.7	10.5	NR	5.4	9.9	NR	0.4	3.9	NR
	Karnataka	27.4	28.1	NR	3.2	6.2	NR	0.0	1.4	NR	0.5	0.9	NR
Sugarcane	Maharashtra	30.8	38.8	NR	15.9	15.4	NR	20.2	18.8	NR	1.4	0.6	NR
	U.P.	34.8	40.8	NR	7.3	13.7	NR	8.9	14.2	NR	1.6	3.2	NR

Economics of Crop Production in Selected Developed and Under-developed Villages as Revealed through Farm Level Survey

Although the above economic analysis of the cost of cultivation data broadly indicated the nature and direction of cost and profit structures in the cultivation of principal crops in various states, it concealed many micro level farm characteristics. Also, it did not reflect on the economics of crop cultivation in various size groups of farms. It was, therefore, thought appropriate to analyse the farm level data, collected with the help of pre-structured schedules for the purpose in the year 1990-91. Two relatively developed and two relatively under developed villages in each of the two selected districts from four selected states of Andhra Pradesh, Bihar, Haryana and Rajasthan were studied. Our results (Tables 2.20 to 2.22) indicate that cropping patterns as well as profitability of crop cultivation varied not only between developed and under-developed regions, but also between different size groups of farms in various states.

ANDHRA PRADESH

It may be seen from Table 2.20 that in Chukkalur village of Anantapur District (which is a relatively backward district, marginal and small farmers cultivated during kharif season mainly rice, groundnut, pulses and jowar, while large farmers cultivated also cotton crop in addition. In the rabi season, marginal farmers were found to grow only groundnut, while small, medium and large farmers cultivated mainly rice on irrigated lands. In the relatively under developed village of Chenam-pally, however, rice, groundnut and pulses formed the cropping pattern in all categories of farms during kharif season, while rice appeared to be the only crop being cultivated on small scale (in irrigated land) in the rabi season by all size groups of farms. In the relatively developed district of West Godavari, rice was grown during both Kharif and rabi seasons by all size groups of farms in both Ponangi and Vatlure village. Tables 2.21 and 2.22 show the profitability of crop cultivation in selected villages of Anantapur and West Godavari Districts in Andhra Pradesh. It may be seen from the tables that gross as well net returns per hectare of net cultivated area and net returns per farm varied widely from region to region. The difference was particularly glaring between the under developed villages of Anantapur and developed villages of West Godavari. The annual gross returns per hectare of net cultivated

TABLE 2.20 : Cropping Pattern in Selected villages by Size Class (% Area Under Various Crops as of 1990-91)

State	District	Village	Cropping pattern (% area)							
			Marginal		Small		Medium		Large	
(1)	(2)	(3)	(4)		(5)		(6)		(7)	
Andhra Pradesh	Anantapur	Chukkalur	G.Nut	48	G.Nut	45	Paddy	45	Pulses	36
			Paddy	31	Paddy	34	G.Nut	35	G.Nut	24
			Pulses	21	Pulses	13	Cotton	20	Cotton	23
					Jowar	8			Jowar	7
		Chanampally	Paddy	59	G.Nut	40	G.Nut	40	G.Nut	50
			G.Nut	37	Pulses	35	Paddy	40	Paddy	40
			Pulses	3	Paddy	18	Pulses	20	Pulses	10
	West-Godavari	Ponangi	Paddy	98	Paddy	90	Paddy	85	Paddy	90
			Veg.	2	S.Cane	10	S.Cane	15	S.Cane	10
		Vatlure	Paddy	89	Paddy	82	Paddy	90	Paddy	90
			Veg.	11	S.Cane	18	S.Cane	10	S.Cane	10
Bihar	Bhagalpur	Miranand	Paddy	40	Paddy	52	Paddy	43	Paddy	50
			Wheat	15	S.Cane	30	S.Cane	35	S.Cane	31
			S.Cane	10	Pulses	8	Pulses	20	Pulses	19
			Veg.	12	Veg.	10	Raicha	2		
			Pulses	8						

(Contd.)

TABLE 2.20 : (*Contd.*)

(1)	(2)	(3)	(4)		(5)		(6)		(7)	
		Mathurapur	Paddy	48	Paddy	31	S.Cane	25	S.Cane	30
			Wheat	26	S.Cane	20	Wheat	19	Paddy	22
			Pulses	20	Maize	13	Paddy	16	Maize	15
			Veg.	6	Wheat	13	Maize	8	Wheat	19
					Pulses	13	Veg.	16	Pulses	14
					Veg.	6	Pulses	11		
					Raicha	4	Raicha	5		
	Patna	Simra	Paddy	62	Paddy	56	Paddy	52	Paddy	40
			Wheat	32	Wheat	30	Wheat	19	S.Cane	30
			Veg.	6	Veg.	14	Pulses	14	Wheat	13
							Raicha	14	Gram	10
									Linseed	7
		Niyamatchak	Paddy	60	Paddy	55	Paddy	50	Paddy	42
			Wheat	30	Wheat	25	Wheat	20	Wheat	30
			Pulses	10	Pulses	15	Linseed	20	Pulses	25
					Linseed	10	Pulses	10	S.Cane	2
Haryana	Bhiwani	Chandabass-I	Bajra	50	Mustard	50	Cotton	20	Berseem	35
			Mustard	50	Bajra	36	Bajra	20	Cotton	35
					S.Cane	14	S.Cane	10	Bajra	15
							Mustard	10	S.Cane	15
							Gram	10		
							Berseem	20		
		Chandabass-II	Bajra	30	Cotton	45	Bajra	45	Bajra	45
			Berseem	30	Bajra	25	Berseem	25	Berseem	25
			Cotton	20	Berseem	30	Cotton	20	Cotton	25
			Paddy	20			S.Cane	10	S.Cane	5

	Karnal	Gagsina	Wheat	42	Paddy	40	Paddy	42	Paddy	46
			Paddy	35	Wheat	40	Wheat	40	Wheat	36
			S.Cane	15	S.Cane	10	S.Cane	12	S.Cane	19
			Berseem	8	Gram	10	Berseem	6		
		Munak	Wheat	50	Wheat	48	Wheat	40	Wheat	50
			Paddy	30	Paddy	36	Paddy	38	Paddy	42
			S.Cane	20	S.Cane	16	S.Cane	18	S.Cane	8
					Gram	5				
Rajasthan	Bikaner	Kyd.	Wheat	30	Wheat	30	Bajra	22	Wheat	30
			Cotton	20	Cotton	15	Wheat	20	Bajra	20
			Jowar	10	Mustard	25	Mustard	20	Guar	20
			Gram	20	Jowar	10	Cotton	10	Pulses	20
			Mustard	20	Gram	20	Guar	18	Cotton	10
							Gram	10		
	Sri Ganga Nagar	Netewala	Cotton	30	Cotton	25	Wheat	25	Wheat	25
			Wheat	20	Wheat	30	Cotton	20	Bajra	25
			Mustard	20	Guar	14	Guar	20	Cotton	15
			Guar	10	Mustard	11	Bajra	10	Gram	20
			Gram	10	Gram	10	Gram	10	Guar	10
			Jowar	10	Jowar	10			Mustard	5
		Amarsar	Cotton	10	Cotton	15	Bajra	15	Bajra	15
			Gram	40	Mustard	40	Wheat	42	Wheat	35
			Mustard	30	Gram	30	Guar	10	Guar	15
			Bajra	20	Bajra	20	Gram	18	Gram	20
							Mustard	15	Mustard	15

TABLE 2.21 : Economics of Crop Cultivation in Selected districts (As of 1990-91)

State	District	Statistics	Annual Gross returns per ha. of net cultivated area (Rs.)	Operational cost (Rs./ha.)	Annual net returns per ha. of net cultivated area (Rs.)	Annual net returns per farm (Rs.)	Output input ratio
Andhra Pradesh	Anantapur	$\bar{X}$	2900.8	1276.6	1615.9	7336.0	2.3
		SD	757.6	276.6	489.7	6977.1	0.2
		CV%	26.1	21.7	30.3	95.1	8.7
	West Godavari	$\bar{X}$	22569.1	10632.5	11936.7	55333.8	2.1
		SD	3238.9	1576.9	1705.6	55532.9	0.05
		CV%	14.4	14.8	14.3	100.4	2.5
Bihar	Bhagalpur	$\bar{X}$	7925.5	3514.2	4411.3	17281.0	2.3
		SD	1448.0	993.5	521.8	15603.4	0.3
		CV%	18.3	28.3	11.8	90.3	12.2
	Patna	$\bar{X}$	9474.6	3302.0	5705.4	28467.3	3.0
		SD	1971.5	467.3	2065.4	24421.1	0.3
		CV%	20.8	14.2	36.2	85.8	8.4
Haryana	Bhiwani	$\bar{X}$	5726.5	2256.5	3470.8	14682.8	2.8
		SD	835.2	849.9	303.0	12692.3	0.6
		CV%	14.6	37.7	8.7	86.4	21.7
	Karnal	$\bar{X}$	11347.8	3983.4	7364.3	37030.0	2.9
		SD	3093.6	1332.1	1792.5	32103.7	0.2
		CV%	27.2	33.4	24.3	86.7	8.0
Rajasthan	Sriganganagar	$\bar{X}$	8150.5	2585.8	5614.8	27019.1	2.9
		SD	5018.3	1147.1	3851.5	34735.5	0.7
		CV%	61.6	44.4	68.6	128.6	22.2

TABLE 2 22 : Economics of Crop Cultivation in Selected Villages (As of 1990-91)

(*Rs./ha.*)

State	District	Village	Size class (Ha)	Annual gross returns per ha. of net cultivated area	Operational Cost/(Rs./ha.)	Annual net returns per ha. of net cultivated area	Annual net returns per Farm (Rs.)	Output input ratio
(1)	(2)	(3)	(4)	(5)	(6)	(7)	(8)	(9)
Andhra Pradesh	Anantapur	Chukkalur	Marginal (.32)	3611	1527	2048	655	2.36
			Small (1.02)	3750	1628	2092	2134	2.30
			Medium (6.07)	3710	1484	2226	13512	2.50
			Large (10.02)	3500	1470	2030	20544	2.38
		Chenam-pally	Marginal (.35)	2028	1011	1017	356	2.00
			Small (1.42)	2492	1278	1214	1724	1.95
			Medium (6.15)	2080	871	1209	7435	2.38
			Large (11.3)	2035	944	1091	12328	2.16
	West-Godavari	Ponangi	Marginal (0.51)	24050	10726	13324	6795	2.24
			Small (1.3)	26378	12662	13716	17831	2.08
			Medium (5.10)	25208	11848	13360	68136	2.13
			Large (12.0)	26912	12920	13992	167904	2.08
		Vatlure	Marginal (0.36)	18565	8911	9654	3475	2.08
			Small (1.08)	19270	9057	10213	11030	2.13
			Medium (4.89)	21120	10138	10982	53702	2.08
			Large (11.10)	19050	8798	10252	113797	2.17

(*Contd.*)

TABLE 2.22 : (*Contd.*)

(1)	(2)	(3)	(4)	(5)	(6)	(7)	(8)	(9)
Bihar	Bhagalpur	Miranand	Marginal (0.30)	7600	3060	4540	1362	2.48
			Small (1.51)	10260	5030	5230	7897	2.04
			Medium (4.46)	10050	4925	5125	22858	2.04
			Large (10.50)	8028	4010	4018	42189	2.00
		Mathurapur	Marginal (0.29)	5588	1956	3632	1053	2.86
			Small (1.07)	7250	2912	4338	4642	2.49
			Medium (4.02)	7428	2971	4457	17917	2.50
			Large (10.21)	7200	3250	3950	40330	2.21
	Patna	Simra	Marginal (0.32)	8830	2598	6232	1994	3.40
			Small (1.21)	12262	3802	8460	10237	3.23
			Medium (5.67)	12198	3856	8342	47299	3.16
			Large (10.12)	10571	3958	6613	66924	2.67
		Niyamat-Chak	Marginal (0.40)	6013	2974	3539	1216	2.86
			Small (1.42)	8446	3040	5406	7677	2.78
			Medium (6.56)	9027	3068	5959	39001	2.94
			Large (10.0)	8450	3120	5330	53300	2.70
Haryana	Bhiwani	Chandabass-I (UD)	Marginal (0.49)	4728	1418	3310	1622	3.33
			Small (1.62)	4931	1520	3411	5526	3.24
			Medium (5.7)	5056	1516	3540	20178	3.33
			Large (10.3)	4930	1525	3405	35753	3.23

		Chandabass-II (D)	Marginal (0.32)	6325	2150	4175	1336	2.94
			Small (1.17)	6922	3498	3424	4006	1.98
			Medium (5.3)	6500	3025	3475	18418	2.15
			Large (10.12)	6420	3400	3026	30623	1.89
	Karnal	Gagsina	Marginal (0.73)	10052	3021	7031	5133	3.33
			Small (1.62)	15226	5633	9593	15541	2.70
			Medium (6.48)	16105	6120	9985	64703	2.63
			Large (10.73)	14278	5282	8996	96527	2.70
		Munak	Marginal (0.61)	8020	2826	5194	3168	2.84
			Small (1.7)	8415	2902	5513	9372	2.90
			Medium (6.17)	10018	3125	6893	41841	3.20
			Large (10.5)	8668	2958	5710	59955	2.93
Rajasthan	Bikaner	Kyd. (Irrigated)	Marginal (0.40)	7651	2850	4801	1920	2.68
			Small (1.42)	8200	3020	5180	7356	2.72
			Medium (6.32)	9210	3192	6018	38034	2.89
			Large (13.40)	9865	3392	6473	86738	2.91
	Sri Ganga Nagar	Netewala (Irrigated) (80%)	Marginal (0.57)	9082	3012	6070	3460	3.02
			Smal (1.46)	13506	3346	10160	14836	4.03
			Medium (4.18)	15002	4210	10792	45111	3.56
			Large (11.3)	14018	4100	9918	112073	3.42
		Amarsar (un-irriga-ted)	Marginal (0.81)	3262	1206	2056	1665	2.70
			Smal (1.94)	3602	1515	1687	4048	2.38
			Medium (6.48)	3820	1768	2052	13297	2.16
			Large (12.15)	3312	1529	1783	21663	2.17

area ranged from Rs. 2,035 in Chenampally (large farms) to Rs. 26,912 in Ponangi (large farms). The annual net returns per hectare varied between Rs. 1,091 in Chenampally to Rs. 13,992 in Ponangi. Both gross and net returns per hectare of net cultivated area were very low in the villages of Anantapur district as compared to those of West Godavari District. In Anantapur, small and medium farms had relatively higher returns per unit of net area than both marginal and large farms mainly because of their more intensive use of land and labour.

But in the relatively developed district of West Godavari where farming was highly capital intensive, large farms in Ponangi village and medium farms in Vatlure village had an edge over all other categories of farms. However, the productivity of small farms were only second in order and were higher than those of marginal and medium farms in Ponangi village and marginal and large farms in Vatlure. Nevertheless, due to relatively low size of holdings, annual net returns per farm were comparatively very low on marginal and small farms in both developed and under-developed villages.

BIHAR

In Bihar, two villages in the district of Patna and two villages in the district of Bhagalpur were studied. It could be seen from Table 2.20 that in Simra village of Patna district, marginal and small size groups of farms cultivated mainly rice during kharif season and wheat during rabi season. But large farmers had substantial acreage under sugarcane which is a standing crop and gram and linseed in the rabi season in addition to wheat. The medium farmers grew mainly rice during kharif and wheat, raicha, gram and other pulses during rabi season. In Niyamatchak village also, a similar cropping pattern was observed in the kharif season by all size groups of farms, while in the rabi season, marginal and small farmers were found to grow linseed, gram and other pulses.

In Miranand village of Bhagalpur district, different size groups of farms followed almost a uniform cropping pattern excepting that some marginal farms in the kharif season grew also maize and vegetables in addition to paddy, sugarcane, gram and other pulses during kharif season and wheat in the rabi season in place of paddy being cultivated by all other size groups of farms. The difference could be due to marginal farmers' preference to opt for relatively low capital intensive wheat crop than rabi paddy (due to capital constraint). In Mathurapur village of Bhagalpur, however, marginal farmers cultivated mainly paddy and vegetables during kharif season and wheat and other pulses (other than gram) in rabi season, while small and medium farmers followed more or less uniform cropping pattern with rice, maize, vegetables (and sugarcane which is a standing crop) during kharif and wheat, gram, raicha and pulses in the rabi season. Large farmers grew mainly paddy, maize, pulses and sugarcane in the kharif season and wheat, gram and pulses in the rabi season. Table 2.22 shows that the annual gross returns per hectare of net cultivated area ranged from Rs. 5,500 in Mathurapur (marginal farms) to Rs. 10,262 in Simra (small farms). Both gross and net returns per hectare were found to be relatively higher on small and medium farms than marginal and large farms in all the four villages of Bhagalpur and Patna districts studied. The net crop income per marginal farm ranged from Rs. 1,053 in Mathurapur (Bhagalpur) to Rs. 1,994 in Simra (Patna) and that of small farms varied from Rs. 4,642 in Mathurapur to Rs. 10,237 in Simra. It is evident from the table that net crop income of marginal and small farms were below the subsistence level in all the four villages of Bihar studied.

HARYANA

Haryana is one of the agriculturally developed states in the country. Nevertheless, there are some pockets in the state where crop productivity is found to be low. Therefore, the relatively developed district of Karnal and relatively under developed district of Bhiwani were selected for the purpose of this study. Besides, the state of Haryana itself was selected for the sake of comparing the economics of crop cultivation of a developed region with that of relatively under developed regions of Bihar, Andhra Pradesh and Rajasthan. In both Chandabass-I and Chandabass-II villages of Bhiwani district, bajra was the main crop, occupying nearly 40 per cent of the total cropped area, followed by cotton, gram, berseem, mustard and sugarcane. It may be seen from Table 2.20 that in Chandabass-I, marginal and small farmers grew mainly bajra, while medium and large farmers had grown cotton (and sugarcane) during kharif season and mustard and berseem in the rabi season. During rabi season, marginal farmers cultivated mustard only. In Chandabass-II however, there was no significant difference in the cropping pattern on different categories of farms, as all size groups of farms excepting large farms cultivated mainly cotton and bajra in the kharif season. However, during rabi season, marginal farms cultivated paddy and berseem, while small, medium and large farms cultivated only berseem, as the remaining cropped area was occupied by a standing crop like sugarcane. In the district of Karnal, paddy and sugarcane formed the cropping pattern in both the village *i.e.*, Gagsina and Munak during kharif season for all categories of farmers. During rabi season, marginal and small farmers of Gagsina cultivated paddy and berseem, while medium and large farmers grew mainly wheat and berseem. In Munak, wheat turned out to be the main crop in the rabi season.

Thus, while in the developed district of Karnal, paddy wheat, sugarcane and berseem formed the cropping patterns, in the relatively backward district of Bhiwani, bajra, cotton, gram, sugarcane and berseem were the main crops grown.

It may be seen from Table 2.22 that gross returns per hectare of net cultivated area ranged from Rs. 4,728 on marginal farms in Chandabass-I (Bhiwani) to Rs. 16,105 on medium farms in Gagsina (Karnal). The net returns per hectare of net cultivated area varied between Rs. 3,026 on large farms in Chandabass-II and Rs. 9,985 on medium farms in Gagsina. The net crop income of marginal farms ranged between Rs. 1,336 in Chandabass-II to Rs. 5,133 in Gagsina and that of small farms ranged between Rs. 4,006 in Chandabass-II and Rs. 15,541 in Gagsina. The output-input ratio varied between 1.89 on large farms in Chandabass-II to 3.33 on marginal and medium farms in Chandabass-I and marginal farms in Gagsina.

RAJASTHAN

In Rajasthan, the relatively developed district of Sriganganagar and relatively under developed district of Bikaner were studied for the purpose of this study. In Bikaner, bajra, guarseed, gram, wheat, mustard, cotton and sesamum were the main crops grown, while in Sriganganagar cotton, wheat, gram, guarseed, mustard and bajra formed the cropping pattern (Table 2.20). There was no significant difference in the cropping pattern of various size groups of farms in Bikaner, as bajra, cotton and guarseed in the kharif season and wheat, mustard, gram and other pulses in the rabi season were generally grown by all size groups of farms. In Sriganganagar, marginal and small farms in Netewala (developed) village grew mainly cotton and guar in kharif season and wheat, mustard and gram in the rabi season; while medium and large farms cultivated mainly paddy, bajra and sugar in the kharif season and wheat, mustard and gram in the rabi season. In Amarsar village (under developed), marginal and small farms grew mainly cotton in the kharif season and gram and mustard in the rabi season. However, medium and large farms cultivated cotton and guar in the kharif season and wheat and gram in the rabi season.

It may be seen from Table 2.22 that gross returns per hectare of net cultivated area varied between Rs. 3,202 on small farms in amarsar and Rs. 15,002 on medium farms in Netewala. The net returns ranged between Rs. 1,783 on large farms in Amarsar and Rs. 10,792 on medium farms in Netewala. The net crop income of marginal farms ranged from Rs. 1,665 in amarsar to Rs. 3,460 in Netewala and that of small farms varied between Rs. 4,048 in Amarsar to Rs. 14,836 in Netewala. While the crop income of marginal and small farms were much below the subsistence level in all the villages studied (excepting small farms of Netewala in Ganganagar district. Infact, Netewala is one of the irrigated villages of Sriganganagar district were small farmers with an average size of 1.46 hectare of land yielded a net crop income of Rs. 14,836). The output-input ratio was higher in Netewala (ranging from 3.02 on marginal farms to 4.03 on small farms) as compared to other villages.

Inference from Macro and Micro Level Analysis of Economics of Crop Farming

It becomes clear from the foregoing (both macro as well as micro level) analysis of economics of crop farming that at the existing levels of yield, income and profit, majority of the farms which are small and marginal would not sustain on long term basis atleast in the dry and rainfed regions. There have been declining trends in factor productivity and rates of profit in the case of most of the cereal crops which act as a disincentive for the farmers to invest more in new technical inputs. In a number of cases (involving both food and commercial crops), the net returns at cost C_2 were found to be negative which indicate that cost effectiveness of the new crop production technology was either non-significant or marginal. Besides, in a number of states including Assam, Bihar, Orissa and Madhya Pradesh, the levels of crop productivity in general were found to be comparatively very low, presumably due to low levels of technology adoption and infrastructural development. Moreover, unequal agrarian structure might have acted as a constraint on development. The results of micro level analysis of crop economics further show that small farmers earned relatively more per unit of net cropped area than marginal and large farmers in most of the villages where farm level survey was conducted. But due to small size of holding, net income from crop farming was not sufficient enough to sustain an average household above poverty line, particularly under unirrigated condition. The situation of marginal farm was worse off, their average net crop income ranging from Rs. 356 in parts of Anantapur District to Rs. 6,795 in parts of West Godavari District of Andhra Pradesh. It is evident that without adequate access to land, irrigation water, technology and other institutional facilities of credit, market and price support, crop based small and marginal farmers would be forced to live below the subsistence level.

Economics of Horticulture

In view of low net returns and income from the cultivation of traditional cereal crops, it is often suggested that diversification in favour of horticulture, sericulture, acquaculture etc. would hold promise for the farmers. The Eighth Five Year Plan specifically mentions about the need for such diversification for sustainable agricultural development. This assumes significance, particularly for dry albeit backward regions which failed to reap the benefits of wheat/rice based green revolution. At the macro level, the economics of horticulture has not been analysed much. Even the National Accounts Statistics give the figures of fruits and vegetables together. India produces nearly 26 million tonnes of fruits and 54 million tonnes of vegetables per annum. Considering the country as a whole, horticultural crops account for 6.3 per cent of the total cropped area. Assuming the triennium ending 1969-70 as the base, the indices of area and production of fruits and vegetables increased to 137.8 and 216.2 respectively in the year 1988-89. The yield index rose to 154.2 in the same year. The total value of fruits and vegetables worked out to Rs. 5,202 crores in 1980-81 and Rs. 5,868 crores in 1986-87 at constant 1980-81 prices. The value of exports of fresh fruits and vegetables increased from Rs. 180.21 crores in 1990-91 to Rs. 456.00 crores in 1994-95 at current prices (Govt. of India, Ministry of Agriculture, 1993). A recent study by IFPRI (1990) shows that there is tremendous potential for expanding exports of fruits and vegetables by India because of comparative advantage in labour intensive, horticultural products. Given appropriate policies and techniques, fruits and vegetables, could add substantially to small farmer's income and employment opportunities. Particularly small farms enjoy adequate economies of scale in the cultivation of vegetable crops. There are a large number of fruit crops which could be profitably grown in various regions of the country. It includes not only major fruit crops like mango, apple, orange, grapes etc., but also olive, hazelnut, almond and walnut in temperate zone and coconut, cashewnut and oil palm in coastal areas which are said to hold lot of promises.

In our study villages, horticulture was not found to have been adopted by farmers on commercial scale, although some large farmers of Anantapur villages were cultivating mango crop quite profitably. In the district of Anantapur, there were about 2,184 hectares of land under mango. Table 2.23 shows the economics of mango cultivation in Penukonda village of Anantapur District.

TABLE 2.23 : Economics of Mango Cultivation in Penukonda Village of Anantapur Malguba

A.	Annual Operational Cost (Rupees per Hectare)	
	(i) Plants	Rs. 5928
	(ii) Labour	Rs. 3705
	(iii) Manure	Rs. 370
	(iv) Chemical Fertilizers	Rs. 988
	(v) Plant Protection	Rs. 370
	(vi) Irrigation	Rs. 350
	Total Operational cost	Rs. 11,711
B.	Annual Average Gross Income from sale of Mango (Rs./ha.)	Rs. 37,050
C.	Net annual Income from sale of Mango (Rs./ha.)	Rs. 25,339

Source: Based on Farm level survey.

It could be seen from Table 2.23 that one hectare of mango (malguba) cultivation yielded a net income of Rs. 25,339 which is difficult to realise from other crops. But there are some practical difficulties in the adoption of mango cultivation by small and marginal farmers. First, there is gestation period of 2 to 3 years before mango plants start giving fruits and during this gestation period, marginal and small farmers need some alternative sources of income for existence. Such a problem does not arise in the case of vegetable crops and other seasonal fruit crops. Besides, the traditional mango crop has a system of bearing fruits in alternate years which acts as a disincentive for the farmers to substitute land under other cereal crops for mango. Moreover, without adequate institutional credit facilities, small and marginal farmers would not be able to invest about Rs. 12,000 in order to realise a net income of Rs. 25,339 per hectare. Furthermore, unless the villages are connected with road and market, middlemen would tend to exploit the fruit and vegetable farmers and reduce their profit margins. Finally, any major increase in the area under fruits and vegetables (which is considered essential atleast in the arid regions) is constrained by the absence of irrigation, adequate high quality seeds and planting material, processing unit and infrastructural support for packaging, storage, cooling and transport. In recent years, the Govt. have encouraged a number of private, corporate and co-orperative efforts, including contract farming for the promotion of fruits and vegetables.

Economics of Sericulture

Sericulture does not strictly fall within the scope of agriculture. Nevertheless, mulberry cultivation is the basis for silk industry and holds lot of promises for generation of income and employment for the small and marginal farmers. It has been reported that sericulture is now practised in about 207 districts in the country, covering 31,073 villages. It provides full time and part time employment to more than 4 million persons in rural areas of which more than 30 per cent belong to scheduled castes and scheduled tribes. It is indeed gratifying to note that there has been a phenomenal growth in the export of silk goods in recent years. India is generally rich in sericigenous fauna which if properly exploited could lead to sericulture revolution in the country. In most of our study villages, sericulture was almost non-existent excepting in Anantapur district of Andhra Pradesh where there has been a remarkable progress in the area under mulberry cultivation recently. There are about 71 thousand acres (28.7 thousand hectares) under sericulture in Anantapur which accounts for 43 per cent of the total area under sericulture in Andhra Pradesh. Table 2.24 indicates the profitability of sericulture in Kondapuram village (Penukonda Mandal) of Anantapur District in Andhra Pradesh. It could be seen from the table that per hectare net return from mulberry cultivation (as of 1990-91) ranged from Rs. 6,000 to Rs. 15,000. The average net return was Rs. 4,562 per hectare. There was large scale inter-farm variations in the returns, due mainly to crop loss for diseases on some farms and here lies the risk of mulberry cultivation by small and marginal farmers. The co-efficient of inter-farm variations in gross returns was as high as 33.1 per cent. In fact, some farmers realised negative returns from mulberry cultivation which acts as a disincentive to the farmers to go for sericulture in a big way. Pests and diseases alone accounts for 20 per cent of crop loss (ESCAP). Unless, there is an effective policy to provide insurance against risk of crop failure and an institutional mechanism for the supply of cross bred disease free layings, small and

TABLE 2.24 : Economics of Sericulture in Kondapuram Village of Anantapur District (Penukonda Mandal)

	Size of Holding (HA.)	Plant Cost/ hect. (Rs.)	Family Labour/hect.		Hired Labour/Hect.		Average input cost/hect. (Rs.)	Value of output/ hect. (Rs.)	Net return/ hect. (Rs.)
			Days	Value (Rs.)	Days	Value (Rs.)			
	0.51	900	320	4400	320	4400	16500	24000	7500
	0.32	800	316	4360	280	4000	16000	10000	–6000
	0.61	1000	320	4400	720	12000	15000	25000	10000
	0.40	800	320	4400	360	4800	18000	16000	–2000
	0.40	1000	320	4400	640	11200	15000	25000	10000
	0.81	1000	320	4400	720	12000	15000	10000	–5000
	1.05	1000	310	4300	478	7180	15000	22000	7000
	2.02	1000	340	4650	580	7800	13000	28000	15000
Mean	0.77	937=50	32=75	4413=75	512=25	7922=50	15437=50	20000=00	4562=50
SD	0.53	85=70	8=00	95=25	166=76	3203=60	1356=41	6614=38	
CV%	68.73	9.14	2.49	2.16	32.56	40.44	8.79	33.07	

marginal farmers may not easily be motivated to switch over to mulberry cultivation. Besides, institutional credit and marketing facilities as well as mobile disinfection units need to be strengthened. Moreover, adoption of improved water management techniques and development of watersheds would be necessary. However, in view of its high employment and income generating capacity, all out efforts should be made for wider adoption of mulberry cultivation wherever feasible.

Economics of Livestock Enterprises

Livestock rearing is considered to be one of the important means of livelihood for many farm households. Considering the country as a whole, the gross value of livestock output was estimated at Rs. 31,513 crores in 1988-89 (National Accounts Statistics, 1991, Government of India). After deducing the feedcost of Rs. 14,929 crores, the net value of livestock output was estimated to be Rs. 16,584 crores. This worked out to be about 15.8 per cent of the gross value added from agriculture and allied activities. However, relatively backward regions of the country have yet to witness white revolution or for that matter significant progress in milk production. An attempt was made in this study to analyse the economics of dairy farming in both developed and underdeveloped villages of our selected districts in a comparative framework.

It could be seen from Table 2.25 that average number of milch animals varied from 1 to 5 in various categories of farms in different selected regions of the country. There is no clear pattern of animal husbandry by various size groups of farms in developed and under developed districts/villages. Nevertheless, while in the relatively developed district of West Godavari, marginal and small farms were found to maintain relatively higher number of milch animals than the large farmers, in the dry, albeit under developed district of Anantapur, the average number of animals maintained by marginal and small farms were comparatively less. However, in both West Godavari and Anantapur Districts, marginal and small farmers were found to maintain milch animals mostly on lease/milk sharing basis. In such cases, 50 per cent of the milk were to be paid to the cattle owner (who were mainly from large farming community), while the entire cost of maintenance was to be borne by the tenants. In the remaining study areas, in the districts of Patna, Bhagalpur, Karnal, Bhiwani, Sriganganagar and Bikaner, no such incidence of leasing of animals was reported.

TABLE 2.25 : Economics of Dairy Farming

State	District	Village	Size Group of Farm (ha.)	Average No. of Milch Animals	Fodder & feed cost per animal (Rs.)	Labour cost per animal (Rs.)	Gross return per animal (Rs.)	Rent paid per animal (Rs.)	Net returns per animal (Rs.)	Net return per farm (Rs.)
(1)	(2)	(3)	(4)	(5)	(6)	(7)	(8)	(9)	(10)	(11)
Andhra Pradesh	West-Godavari	Ponangi	Marginal (0.51)	4	1250	850	5800	2900	800	3200
			Small (1.3)	3	1480	1200	6250	3125	445	1335
			Medium & Large (7.5)	1	2000	1260	6500	—	3240	3240
		Vatlure	Marginal (0.4)	3	1040	900	4775	2387	448	2700
			Small (1.04)	2	1100	1050	5800	2900	750	1500
			Medium & Large (8.2)	1	1330	1250	6200	—	3620	3620
		Chukka-lore	Marginal (0.32)	2	1000	1020	4620	2325	305	610
			Small (1.02)	2	1200	1070	4880	2440	170	340
			Medium & Large (8.4)	4	1250	1280	5400	—	2870	11480
Bihar	Patna	Simra	Marginal (0.32)	2	1150	1030	4850	—	2670	5340
		14.0	Small (1.21)	3	1206	777	4900	—	2917	8751
		26.0	Medium & Large (8.2)	3	1886	1210	5080	—	2984	8952

(Contd.)

TABLE 2.25 : (*Contd.*)

(1)	(2)	(3)	(4)	(5)	(6)	(7)	(8)	(9)	(10)	(11)
		Niyamat-Chak	Marginal (0.40)	2	1250	1100	4650	—	2300	4600
			Small (1.42)	2	1500	1080	5080	—	2500	5000
			Medium & Large (8.7)	3	1710	1200	5825	—	2915	8745
	Bhagalpur	Miranand	Marginal (0.35)	1	1200	1060	4650	—	2390	2390
			Small (1.51)	—	—	—	—	—	—	—
			Medium & Large (7.7)	1	1800	1250	5800	—	2750	2750
Haryana	Karnal	Gagsina	Marginal (0.73)	2	2190	1078	6970	—	3702	7404
			Small (1.62)	3	3046	1408	7850	—	3396	10188
			Medium & Large (9.2)	8	3735	1440	9600	—	4425	35400
		Munak	Marginal (0.21)	2	2225	990	5840	—	2625	5250
			Small (1.70)	2	2530	1250	8060	—	4280	8560
			Medium & Large (8.7)	4	2880	1326	8890	—	4684	18736
	Bhiwani	Chanda-bass-I	Marginal (0.49)	2	1956	922	5556	—	2678	5356
			Small (1.69)	2	2795	1738	6880	—	2347	4694
			Medium & Large (9.2)	3	2138	1512	7038	—	3388	10164

		Chanda-bass-II	Marginal (0.73)	2	1450	992	5050	—	2608	5216
			Small (1.62)	2	1800	1400	5800	—	2600	5200
			Medium & Large (8.7)	4	2050	1450	6250	—	2750	11000
Rajasthan	Sri Ganga-nagar	Netewala	Marginal (0.40)	2	2100	1250	6500	—	3150	6300
			Small (1.42)	3	2650	1400	8100	—	4050	12150
			Medium & Large (10.0)	6	3000	1500	8600	—	4100	24600
		Amarsar	Marginal (0.57)	2	1742	850	4700	—	2108	4216
			Small (1.46)	2	1820	1230	4850	—	1800	6074
			Medium & Large (8.6)	3	2458	1162	6500	—	2880	8640
	Bikaner	Kyd.	Marginal (0.81)	2	2033	768	5000	—	2199	4398
			Small (1.94)	2	2050	900	5042	—	2092	4184
			Medium & Large (10.8)	5	2612	1118	5890	—	2160	10800

TABLE 2.26 : Co-Efficients of Factors Influencing Milk output per Animal

State	Constant Intercept	Farm size (acres)	Expenditure on Fodder & Feed	Expenditure on labour and maintenance	R^2	Adjusted R^2
Andhra Pradesh	3264.134	6.850	1.499	0.2937	0.598	0.356
		(36.601)	(0.8627)	(2.1462)		
Bihar	3991.031	46.988	0.9618	–0.458	0.971	0.950
		(20.688)	(0.5501)	(0.5072)		
Haryana	2080.704	–0.0549	2.035*	0.0208	0.815	0.746
		(0.5347)	(0.5223)	(0.9894)		
Rajasthan	–2709.872	–34.1244	3.1344*	1.906*		
		(18.196)	(0.5504)	(0.694)		
Overall	1733.038	2.063	1.414*	1.440**	0.781	0.769
		(0.664)	(0.1980)	(0.580)		

* Statistically significant at 1 per cent level.
** Statistically significant at 5 per cent level.

It could be further seen from Table 2.26 that there is a positive relationship between farm size and gross as well as net returns per milch animal in most places which could be explained by large farmer's access to improve feed for animals including green fodder, dry fodder and concentrates. The relationship between expenditure on fodder and milk output per animal also was found to be positive and statistically significant. The marginal and small farmers do have neither enough land for the purpose of growing fodder nor adequate capacity to buy fodder and concentrates from others. Generally speaking they depend on common grazing lands and pastures and remain satisfied with low milk yields and returns. Besides, in places where public animal health care facilities are not available, small and marginal farmers find it difficult to maintain cross bred, albeit high yielding milch animals due to their more proneness to diseases. Particularly, reproductive disorders in the milch animals often cause considerable economic loss which small and marginal farmers find hard to bear. Moreover, in under-developed villages, due to lack of market, marginal farmers' who are dependent on the large farmers for employment and credit, tend to sell their milk produce to their patrons at relatively lower prices. In addition, it is borne out by the table that in the districts of West Godavari and Anantapur, small and marginal farmers who maintain milch cattle mainly on lease or milk sharing basis, earn lower net return per animal also because they part away with 50 per cent of the milk produce to the cattle owners.

Furthermore, it needs to be emphasised here that cattle rearing is only a part time or supplementary occupation for average Indian farmers and its viability should be examined as part of farming system and not so much as an independent occupation. It is true that experiences of other countries and some research studies in India show that cattle rearing by itself could be a viable and sustainable occupation (Chaudhary and Singh, 1990), generating lot of income through sale of milk, dung and manure, calves and other animal stock. But the initial capital cost for the purchase of milch stock, construction of dairy sheds and availability of land for the production and maintenance of fodder would prevent many marginal and small farmers to undertake independent dairy farming due to lack of their own investible surplus as well as inadequate institutional credit facilities prevailing in greater part of the country. Besides, lack of organised marketing and processing facilities stand in the way of dairy development programme in general. Moreover, there is an opportunity cost involved for many marginal farmers who could not hire out labour outside in order to maintain cattle in the villages of their residence. A recent study (Maji *et al*, 1995) shows that even an integrated crop and livestock farming system is not profitable for marginal farmers if they have to buy fodder and other concentrates for animals. The present system of milch animal rearing by them is reported to sustain mainly because of free or subsidised fodder supply from their village patrons and availability of grazing and pasture lands. The erosion of common grazing lands in recent years has further reduced the small farmers' ability and incentive to maintain milch cattle on profitable basis. This is true not only for cattle rearing, but also for goat and sheep rearing and poultry by small and marginal farmers.

Economics of Fish Cultivation

Acquaculture is considered to be one of the important ways of occupational diversification and source of additional income and employment for the farmers. It is particularly emphasised at the research station level that integrated farming system involving paddy cum fish culture, fish cum horticulture, fish-duck integration etc., can become an effective source of additional income generation for the country. Unfortunately, however, this potential remains unutilised thus far. The share of fish production in the gross domestic product of the country remained more or less constant over time at 0.8 per cent during 1980-81 to 1990-91 (at constant prices). Also, the per-capita net availability of fish remained constant during this period at about 9 grams per day.

As per the latest statistics (Government of India, Ministry of Agriculture, 1995), marine fisheries contribute nearly 57.4 per cent of the total fish production in the country. As of 1993-94, there were about 4681 thousand tonnes of fish production, of which 2688 thousand were contributed by marine fisheries and 1993 thousand tonnes were contributed by inland fisheries. Of the total marine fish production, 72.3 per cent were shared by the states of Gujarat (19.8 per cent), Kerala (18.6 per cent), Maharashtra (18.6 per cent) and Tamil Nadu (15.3 per cent). As regards inland fisheries, the state of West Bengal contributed the highest share of 33.9 per cent, followed by Bihar (11.2 per cent), Andhra Pradesh (9.1 per cent), Uttar Pradesh (6.8 per cent), Tamil Nadu (6.7 per cent) and Orissa (5.3 per cent). The shares of all other states ranged from 0.1 per cent (Arunachal Pradesh) to 4.1 per cent (Madhya Pradesh). However, various micro level studies reveal that production of both inland and marine fish could be sufficiently raised through appropriate technological changes, credit, market and other infra-structures. In the case of Andhra Pradesh alone, it has been estimated that about 17,000 hectares of Brackish water areas are suitable for immediate development of fish culture. The Brackish water prawn culture in coastal Andhra Pradesh has proved to be a very profitable occupation which attracts not only the local farmers, but also outside entrepreneurs. In this context however, it may be necessary to safeguard the interests of local small and marginal farmers through development of suitable co-operatives.

In most of our study villages, fish cultivation was more or less non-existent. But in Ponangi village of West Godavari District, nearly 5 large and medium farmers had adopted fish culture which generated an average income of Rs. 39,040 per hectare. Table 2.27 indicates the average input-output matrix of fish production in the said village.

TABLE 2.27 : Economics of Fish Cultivation in Ponangi Village of West Godavari District

A.	Annual Average Cost per Hectare (Rs.) of Fish Tank :	
	(i) Seed Value	Rs. 8,675
	(ii) Labour	Rs. 2,440
	(iii) Other expenses including feed, interest on loan etc.,	Rs. 32,560
	Total operational cost	Rs. 43,675
B.	Annual Gross Income per Hectare of Fish Tank (Rs.)	Rs. 82,715
C.	Annual Net Income per Hectare of Fish Tank (Rs.)	Rs. 39,040

Source: Based on farm level survey.

It is clear from the table that per hectare annual income from fish production is much higher than that of crop production and fish culture may appear to be a viable proposition for small and marginal farmers. But small and marginal farmer's inadequate access to land and institutional market and credit facilities stand in the way of their adopting fish cultivation on commercial scale. Besides, a change from crop cultivation to fish culture involves certain risks and uncertainties which most marginal and small farmers tend to avert due to their low capacity to bear risk of production/income loss arising out of fish diseases, theft etc. Moreover, any such change may tend to convert good crop land into fish tanks and it may be necessary to ensure that only less productive lands are used for fish culture keeping in view the greater social interest.

REFERENCES

Chaudhury, A.L., and I.J. Singh (1990). 'Planning for Agricultural Development in Trans Gangetic Plains Regions (Zone-VI)', HAU, Hissar, Oct., 1990 (occasional paper).

ESCAP as reported by V.L. Chopra, Former Director-General, ICAR in his speech at the Special General Meeting of the ICAR at Hyderabad, August 30, 1992.

George, M.A., N.J. Kurien and Mohan, C. Chandra, 'Factor Shares in Indian Agriculture', *Indian Journal of Agricultural Economics*, Vol. XXXVIII, No. 3, July-Sept. 1983.

Government of India, Ministry of Agriculture (1991). Cost of Cultivation of (1993) principal crops in India, DE&S, New Delhi.

Government of India (1991). National Accounts and Statistics, CSO, New Delhi.

Government of India, Ministry of Agriculture (1993), Agricultural Statistics at a Glance, DE&S, New Delhi.

Haque, T. (1992). Economics of Agriculture in Backward Regions with special Reference of Viability and Sustainability of Small and Marginal Farms, NIRD (mimeo).

The Hindu Daily, Hyderabad, Feb. 7, 1994.

IFPRI (1990). Nural Islam, 'Horticultural Exports of Developing Countries : Past Performances, Future Prospects and Policy Issues', IFPRI, Research Report No. 80, Washington D.C., 1990.

Maji, C.C. T. Haque and A. Bhattacharya, 'Small Farms, Employment and Surplus Generation', NCAP, Policy Paper No. 5, New Delhi, 1995.

Ramana, D.V., 'Economics of Sericulture and Silk Industry in India', Deep & Deep Publications, New Delhi, 1987.

3

VIABILITY AND SUSTAINABILITY OF MARGINAL AND SMALL FARMS

Whether marginal and small farms are economically viable or not, could be examined by analysing the profit and income levels of farms under varying crop mix or farming systems. The profit and income variables in turn depend on a number of factors including agro-climatic conditions, technology, price, cropping/farming patterns, size of holding and availability of infra-structure/institutional facilities of credit, irrigation, market etc. But the basic question that would arise here is how do we define viability? In fact, the term 'viability' could be defined in several ways. First, as long as any crop or farm activity shows net positive returns, the said farm could be called economically viable, irrespective of its size (criterion-1). Second, in a given situation, a farm which generates sufficient income not only for meeting the consumption needs of an average houschold, but also for saving and investment for future development would be called economically viable (criterion-2). Besides, if such viability is found to have been established over time, this could also be accepted as sustainable on long-term basis. However, the term 'Sustainability' has some normative meanings too. Whatever has been found sustainable in the past, may not necessarily sustain in the future, due to the failure to follow certain norms of agricultural practices. For instance, a failure to maintain agro-climatic environment due to soil degradation, deforestation, population growth, technological stagnation and market imperfection, may tend to convert the existing viable farms into non-viable and non-sustainable ones over a period of time. Therefore, the question of viability and sustainability of marginal and small farms should have to be examined in the light of these facts.

Viability and Sustainability of Farms on the Basis of Crop Cultivation only

It may be seen from Table 3.1 that marginal farms could no where (including the developed districts of West Godavari, Karnal, Patna and Sriganganagar) generate sufficient income from crop farming alone to stay above poverty line. Marginal farms were found to be non-viable according to both criterion-1 and criterion-2. Per hectare net returns of marginal farms were generally lower than those of other categories of farms almost everywhere excepting Chukkalur (Anantapur), Miranand (Bhagalpur) and Chandabass-II (Bhiwani) where net returns per hectare of marginal farms were slightly higher than that of large farms. Of course, if we judge viability on the basis of efficiency measured in terms of factor productivity or profit function, then all size groups of farms including marginal ones were found to be viable in our study villages. In this case, the form of simplified, albeit normalised profit function would be as follows :

$$TT = \beta O \, (W/P)^{\beta 1} \, (K/P)^{\beta 2}$$

Where TT = Total value of crop output minus the total cost of variable inputs in rupees;

W/P = Normalised wage rate of human labour in rupees.

K/P = Weighted price of capital inputs including chemical fertilizers, insecticides, machine labour etc.

However, it could be seen from Table 3.2 that fertilizer use efficiency of small and marginal farms was higher than that of large farms in a number of villages (particularly in irrigated areas), while large farms were found to be better in terms of human labour use efficiency in general. However, in a labour surplus agricultural economy like ours, it is not merely the marginal value product of labour, but also the overall employment of labour per unit of land or capital which is important. In this context, small and marginal farms were found to use more labour per unit of land. Nevertheless, judged on the basis of overall output-input ratio or factor productivity, small and medium size groups of farms were found to be relatively more efficient than large and marginal farms in most of the villages studied.

Judged by the second criterion, small farms could generate crop income above the subsistence level only in the relatively developed villages of Ponangi, Vatlure (West Godavari), Gagsina (Karnal) and

TABLE 3.1 : Income-Savings Profile of various Size Groups of Farms

State	District	Village	Size Group	Average crop income (Rs.)	Average total income from all sources (Rs.)	Minimum level of consumption exp. needed to be above pov. line (Rs.)	Average saving from crop income (Rs.)	Average saving from all sources (Rs.)
(1)	(2)	(3)	(4)	(5)	(6)	(7)	(8)	(9)
Andhra Pradesh	Anantapur	Chukkalur	Marginal	655	4630	11000	–10345	–6370
			Small	2134	5270	11000	–8866	–5730
			Medium	13512	18992	11000	2512	7992
			Large	20544	26268	11000	9544	15268
		Chenam-pally	Marginal	356	4025	11000	–10644	–6975
			Small	1724	4628	11000	–9276	–6372
			Medium	7435	11035	11000	–3565	35
			Large	12328	18348	11000	1328	7348
	West-Godavari	Ponangi	Marginal	6795	11079	11000	–4205	79
			Small	17831	20386	11000	6831	9386
			Medium	68136	78392	11000	57136	67392
			Large	167904	193524	11000	156904	182524
		Vatlure	Marginal	3475	9098	11000	–7525	–1902
			Small	11030	14530	11000	30	3530
			Medium	53702	58332	11000	42702	47322
			Large	113797	124047	11000	102797	113047

(Contd.)

TABLE 3.1 : (*Contd.*)

(1)	(2)	(3)	(4)	(5)	(6)	(7)	(8)	(9)
Bihar	Bhagalpur	Miranand	Marginal	1362	4612	11000	–9638	–6388
			Small	7897	9886	11000	–3103	–1114
			Medium	22858	25420	11000	11858	14420
			Large	42189	44939	11000	31189	33939
		Mathurapur	Marginal	1053	4479	11000	–9947	–6521
			Small	4642	7644	11000	–6358	–3356
			Medium	17917	20417	11000	6917	9417
			Large	40330	41280	11000	29330	30280
	Patna	Simra	Marginal	1994	10596	11000	–9006	–404
			Small	10237	18988	11000	–763	7988
			Medium	47299	56251	11000	—	—
			Large	66924	75876	11000	55924	64876
		Niyamatchak	Marginal	1216	8906	11000	–9784	–2094
			Small	7677	12677	11000	—	—
			Medium	39091	47836	11000	—	—
			Large	53300	62045	11000	—	—
Haryana	Bhiwani	Chandabass-I	Marginal	1622	12378	11000	–9378	1378
			Small	5526	11032	11000	—	—
			Medium	20178	30342	11000	—	—
			Large	35753	45918	11000	—	—
		Chandabass-II	Marginal	1336	12052	11000	–9664	1052
			Small	4006	14035	11000	—	—
			Medium	18418	—	11000	—	—
			Large	30623	—	11000	—	—

	Karnal	Gagsina	Marginal	5133	18162	11000	–5867	7167
			Small	15541	25729	11000	—	—
			Medium	64703	100103	11000	—	—
			Large	96527	131927	11000	—	—
		Munak	Marginal	3168	13218	11000	–7832	2218
			Small	9372	17932	11000	—	—
			Medium	41841	60577	11000	—	—
			Large	59955	78691	11000	—	—
Rajasthan	Bikaner	Kyd.	Marginal	1920	9018	11000	–9080	–1982
			Small	7356	11540	11000	3644	540
			Medium	38034	48834	11000	27034	37834
			Large	86738	98788	11000	75738	87788
	Sri Ganga-nagar	Netewala	Marginal	3460	12676	11000	–7540	1676
			Small	14836	26986	11000	3836	15986
			Medium	45111	69711	11000	34111	58711
			Large	112073	136673	11000	101073	125673
		Amarsar	Marginal	1665	8581	11000	–9335	–2419
			Small	4048	10122	11000	–6952	–878
			Medium	13297	21937	11000	2297	10937
			Large	21663	30303	11000	10663	19303

TABLE 3.2 : Elasticity Co-Efficients of Factors Influencing Crop Productivity

State	District	Village	Size Group	Fertiliser		Human Labour		R_2
				β	SE	β	SE	
(1)	(2)	(3)	(4)	(5)	(6)	(7)	(8)	(9)
Andhra Pradesh	Anantpur	Chukkalur	Marginal	—	—	0.024	0.028	0.13
			Small	0.036*	0.014	0.0512	0.091	0.36
			Medium & Large	0.041	0.042	0.0312**	0.010	0.42
		Chenam-Pally	Marginal	—	—	0.0501	0.092	0.06
			Small	0.431*	0.146	0.071	0.051	0.43
			Medium & Large	0.016	0.022	0.073**	0.021	0.42
	West-Godavari	Ponangi	Marginal	0.073*	0.031	–0.191	0.263	0.52
			Small	0.092***	0.036	–0.023	0.105	0.56
			Medium & Large	0.031	0.029	–0.078*	0.002	0.34
		Vatlure	Marginal	0.130*	0.044	–0.0745*	0.312	0.66
			Small	0.026*	0.008	–0.831	0.939	0.44
			Medium & Large	0.003*	0.001	0.001	0.001	0.22

Haryana	Bhiwani	Chandabass-I	Marginal	0.003	0.002	−0.452**	0.012	0.38
			Small	0.005*	0.001	−0.391	0.414	0.42
			Medium & Large	0.0241	0.031	−0.031*	0.010	0.40
		Chandabass-II	Marginal	0.001	0.001	−0.111	0.132	0.18
			Small	0.121***	0.023	0.045**	0.011	0.68
			Medium & Large	0.146	0.031	−0.083	0.023	0.46
	Karnal	Gagsina	Marginal	0.0128***	0.008	0.321	0.123	0.58
			Small	0.362	0.021	−0.016	0.023	0.42
			Medium & Large	−0.006	0.005	−0.003*	0.001	0.45
		Munak	Marginal	0.236**	0.062	0.120*	0.052	0.65
			Small	0.31*	0.112	0.003*	0.001	0.68
			Medium & Large	−0.003	−0.003	−0.008	0.012	0.23
Bihar	Bhagalpur	Miranand	Marginal	0.321*	0.0118	−0.112	0.232	0.36
			Small	0.462	0.134	−0.438	0.852	0.43
			Medium & Large	0.032	0.081	0.042*	0.230	0.42
		Mathurapur	Marginal	0.014	0.043	0.002	0.001	0.38
			Small	0.134*	0.046	−0.113	0.020	0.42
			Medium & Large	0.222*	0.082	0.004*	0.002	0.56

(*Contd.*)

TABLE 3.2 : (*Contd.*)

(1)	(2)	(3)	(4)	(5)	(6)	(7)	(8)	(9)
	Patna	Simra	Marginal	0.136*	0.063	0.002	0.004	0.46
			Small	0.431*	0.134	-0.184**	0.023	0.68
			Medium & Large	0.031	0.045	0.003*	0.001	0.42
		Niyamatchak	Marginal	0.032	0.040	–0.019	0.145	0.28
			Small	0.162*	0.004	–0.035*	0.010	0.58
			Medium & Large	0.114*	0.003	0.001	0.011	0.42
Rajasthan	Bikaner	Kyd.	Marginal	0.322	0.356	–0.328	0.141	0.24
			Small	0.234*	0.072	0.006*	0.002	0.58
			Medium & Large	0.136	0.011	–0.0512	0.042	0.19
	Sri Ganga-nagar	Netewala	Marginal	0.004*	0.001	–0.0172	0.023	0.41
			Small	0.016	0.011	–0.033*	0.10	0.38
			Medium & Large	0.138*	0.042	0.014*	0.004	0.62
		Amarsar	Marginal	0.146	0.188	–0.002	0.003	0.32
			Small	0.126*	0.004	–0.023	0.140	0.46
			Medium & Large	0.130	0.246	0.001	0.002	0.28

* Statistically significant at 10 per cent level.
** Statistically significant at 5 per cent level.
*** Statistically significant at 1 per cent level.

Netewala (Sriganganagar). In all other villages, net crop income of small farms were much below the subsistence level of Rs. 11,000. Table 3.1 shows the income-savings profile of various size groups of farms in the selected villages. It could be seen from Table 3.1 that if we assume Rs. 11,000 as the subsistence level of income (which is considered to be the latest poverty line), then income gap (based on own crop income) of marginal farms ranged from Rs. 4205 in Ponangi village of West Godavari to Rs. 10,644 in Chenampally village of Anantapur. In the case of small farms, the income gap ranged between Rs. 3,103 in Miranand village of Bhagalpur to Rs. 9,276 in Chenampally village of Anantapur. In the village of Ponangi, Vatlure (West Godavari), Gagsina (Karnal) and Netewala (Sriganganagar), the savings of small farms from crop income ranged from Rs. 30 in Vatlure to Rs. 6,831 in Ponangi. It is clearly borne out from the table that in the relatively backward village of Chenampally in Anantapur, even the medium farms do not generate enough crop income to stay above subsistence level and large farms had hardly subsistence level of income through crop farming. It is thus clear that by and large, marginal and small farms cannot earn a viable level of income from crop farming alone. This is particularly true for crop farming under dry and rainfed conditions. Table 3.1 further shows that marginal and small farmers depend mainly on other sources of income for their subsistence. These include leasing in land, animal husbandry, poultry, wage paid employment etc. But even when we include income from all other sources, marginal farmers' income levels were found to be below the subsistence level in most of the villages excepting Ponangi in West Godavari, Netewala in Sriganganagar, Chandabass-I in Bhiwani, Gagsina and Munak in Karnal. In other words, marginal farmers who constitute a vest majority of the rural population live below poverty line in greater part of the country. Therefore, according to the second criterion, under existing situation, marginal farms cannot be called viable units even after considering all available sources of income. Similar is the case with small farms in unirrigated villages. Figs. 3.1 to 3.4 show the income gap of marginal farmers below the subsistence level in selected villages, considering own crop income and income from all available sources. Thus, when marginal and small farms are not found viable on the basis of subsistence level of income criterion during a normal agricultural year, their sustainability under adverse circumstances like drought, flood etc. becomes all the more uncertain. Besides, their capacity to sustain gets reduced due to continuous soil degradation, deforestation, erosion of common property resources and ever increasing pressure of population on land.

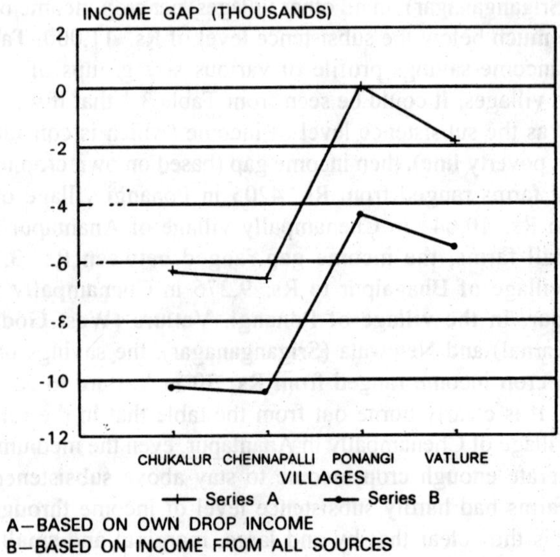

Fig. 3.1 : Income Gap of Marginal Farmers (AP) To Cross Poverty Line

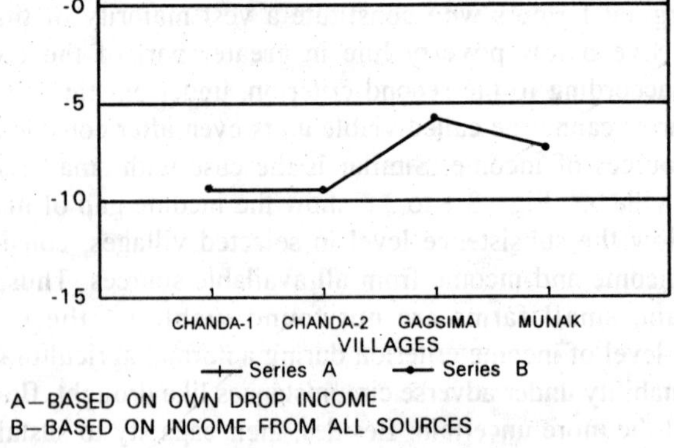

Fig. 3.2 : Income Gap of Marginal Farmers (Har) To Cross Poverty Line

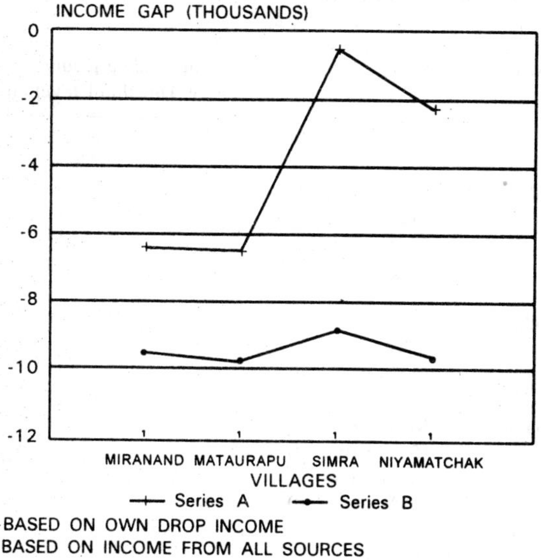

Fig. 3.3 : Income Gap of Marginal Farmers (Bihar) To Cross Poverty Line

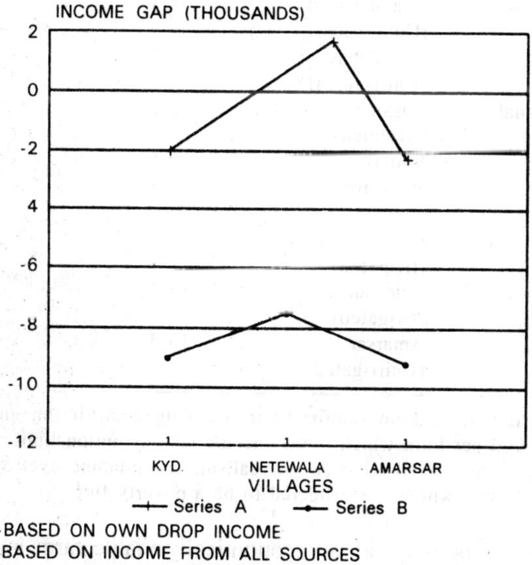

Fig. 3.4 : Income Gap of Marginal Farmers (Raj) To Cross Poverty Line

TABLE 3.3 : Minimum Size of Holding Required to Yield a Thresh Hold Level of Income (Area in Hectare)

State	District	Village	Size of Holding required under alternative Threshold levels of Income	
			I	II
Andhra Pradesh	Anantapur	Chukkalur (partly irrigated)	13.4	5.2
		Chennampally (unirrigated)	23.1	9.1
	West Godavari	Ponangi (irrigated)	2.1	0.8
		Vatlure (70% irrigated)	2.7	1.1
Bihar	Bhagalpur	Miranand (50% irrigated)	5.4	2.1
		Mathurapur (20% irrigated)	6.5	2.6
	Patna	Simra (irrigated)	3.3	1.4
		Niyamatchak (50% irrigated)	5.2	2.1
Haryana	Bhiwani	Chandabass-I (Unirrigated)	8.3	3.3
		Chandabass-II (Unirrigated)	8.2	3.3
	Karnal	Gagsina (Irrigated)	3.0	1.2
		Munak (60% Irrigated)	5.1	2.0
Rajasthan	Bikaner	Kyd. (Irrigated)	5.5	2.2
	Sri Ganga-nagar	Netewala (Irrigated)	2.8	1.1
		Amarsar (Unirrigated)	13.5	5.3

I = refers to size of land required for realising a minimum amount of Rs. 28,000 per farm which is the income tax exemption limit.

II = refers to size of land required for realising a minimum level of income of Rs. 11,000 which is considered to be a poverty line.

Moreover, to be realistic, an economically viable farm is not just the one which generates a little higher than subsistence level of income. A farm household needs income not only for meeting its consumption requirements, but also for facing emergency situation like illness, accident, drought, flood etc. In addition, there has to be some savings for investment for future growth. If we assume that at current prices, Rs. 28,000 per farm is a reasonable level of income for some such sustenance (as this is treated as the income tax exemption limit in urban areas), then even large farms in many dry zones do not generate sufficient income from crop farming alone for the purpose. The villages of Chukkalur and Chenampally in Anantapur and Amarsar in Rajasthan (Table 3.1) provide case instances in this regard. It may be seen from Table 3.3 that minimum size of holding required to generate a threshold level of crop income under existing situation (based on subsistence level of income) ranged from 0.8 hectare in Ponangi village of West Godavari to 9.1 hectare in Chenampally village of Anantapur district. But on the basis of Rs. 28,000 per farm criterion, the minimum required size of land holding ranged from 2.1 hectare in Ponangi village of West Godavari district to 23.1 hectare in Chenampally village of Anantapur district. In fact, the main argument of those who oppose low ceilings on land holdings is that even medium and large farms in dry regions do not generate sufficient income for sustainable development.

However, there is also another side of the picture. Viability as well as sustainability of farms are not static concepts. It varies due to changes in technology, cropping patterns, infrastructural facilities, input-output prices, farming systems, population growth and possibilities of integration between farm and non-farm activities. Besides, even in an agriculturally developed country like Japan where average size of land holding is very small, viability is judged on the basis of efficiency criterion rather than on the basis of level of income. Despite the best available technology and high productivity of land, no Japanese farmers sustain on the basis of crop farming alone and proper integration of farm and non-farm sources of income is considered essential (Kaneda, 1980). Moreover, what is found to be viable to-day, may or may not remain so in future depending on factor-productivity, factor relations and extent of population pressure on land. Infact, given the present trend of population pressure on land in India to continue, most of the marginal households may join the rank of landless labourers soon, which in the absence of sufficient growth of farm and off farm wage paid employment may pose a threat to socio-economic stability in the villages. To the extent, socio-economic environment gets disturbed due to growing landlessness and unemployment, the agrarian system as such looks non-viable and non-sustainable.

Impact of Technological Change

Impact of technological change on viability and sustainability of farms can be examined in various ways. First, viability of farms situated in the technologically developed regions could be viewed against that of technologically under developed regions. Judged by this criterion, it is evident from our farm level data that marginal and small farms in irrigated areas are quite viable in the sense of both efficiency of production and income levels. But dry regions depict a very depressing picture, due to non-availability of dry farming technology and low use of new technical inputs. Due to low level of technological development, even large farms fail to generate minimum necessary income for sustainable development. Second, the agricultural research institutes and universities have generated some new technologies whose productivity potentials are reported to be quite high. For instance, the results of ICAR, National Demonstrations (ICAR, 1986, Haque, 1988) show that there are large technological yield gaps in some of the so-called backward regions (Table 3.4) which if fully utilised, may help enable the marginal and small farms to generate sufficient income above the subsistence level. However, it also depends on crop rotation, cropping patterns and also yield response to technology adoption. Irrigated land capable of producing two or more crops in a year, using recommended level of technology may sustain small and marginal farms on viable basis. But in unirrigated dry regions, farmers are able to grow only one crop with very low yield response to available technology, and therefore, viability (in terms of level of income) of small and marginal farms is a remote possibility there, unless there is a sufficient breakthrough in dry farming technology or a shift from traditional cereal crops to high value crops including fruits, vegetables, sericulture etc. Further, crop or non-crop based diversification would require appropriate infrastructural/organisational structure for credit, marketing, storage and transportation. Moreover, choice of appropriate crop mix or for that matter farming system involving crop production, animal husbandry, horticulture, acquaculture, sericulture etc., would be essential. Our field observations show that there is huge untapped potential for productivity growth in the rural sector and there is no reason why viability of a farm should be judged on the basis of traditional crop farming alone. Technological change, diversification and infrastructural development together tend to reduce the necessary minimum size of land holding required for sustainable development. The small is still beautiful. It needs a little more care and nourishment in terms of availability of appropriate technology, credit, market, storage and insurance. The provision of insurance is particularly important because of the frequent occurrence of drought and flood, the risks of which must be covered for the sustenance of marginal and small farms. Moreover, in view of the rising prices of chemical fertilizers and pesticides and environmental hazards being associated with it, it is essential to develop and promote the judicious use of low cost bio-fertilizers/natural farming by marginal and small farmers.

TABLE 3.4 : Technological/Yield Gaps in Selected Crops

State	Ratio of potential yields to actual yields				
	Rice	Wheat	Maize	Bajra	Jowar
Andhra Pradesh	2.0	—	1.4	1.4	—
Assam	4.6	—	—	—	—
Bihar	3.5	2.1	—	—	—
Gujarat	—	1.7	—	2.1	1.20
Haryana	2.4	1.3	—	1.6	—
Himachal Pradesh	—	—	1.9	—	—
Karnataka	3.1	—	—	—	—
Kerala	—	—	—	—	—
Maharashtra	3.4	2.1	—	9.3	3.4
Orissa	2.6	1.2	—	—	—
Punjab	2.6	—	2.8	—	—
Rajasthan	—	—	—	—	—
Tamil Nadu	1.7	—	—	—	—
Uttar Pradesh	3.0	2.0	2.5	2.0	—
West Bengal	2.2	—	—	—	—
National Average	2.8	1.6	2.1	2.6	4.6

Impact of New Economic Policy of Liberalisation on Small Farm Development

Although it is too early to analyse the impact of new economic policy on small farm development, an ex-ante analysis is both possible and necessary. Unlike socialist countries, agriculture in India has always been in private hands. Therefore, privatisation means very little so far as the ownership and cultivation of land are concerned.

Therefore, the new policy would basically mean only reduction of subsidy and price control. Fortunately, the present subsidy level remains below the stipulated ceiling imposed by WTO. While the reduction of input subsidy may have varying impact on crop output of various regions, depending on the elasticity of crop yield to input use as well as price elasticity of input use (Sharma and Haque), which in turn depend on the levels of development and availability of necessary infrastructural facilities, the removal of price control is being projected as the guarantor of production efficiency and higher income for the farmers.

It is often argued that in the past, input subsidy in India has benefitted mainly the rich farmers and developed regions of the country and hence its withdrawal would release funds which could be utilised for the development of small farms and backward regions in future (Gulati and Bhide, 1993). This is indeed a good idea, but faced with lot of practical difficulties. While the farmer's lobby would put a limit on reduction of input subsidy, it is also not clear how money which could be released due to withdrawal of subsidy be put to use by small farms and backward regions in the absence of an institutional framework for this purpose. Unfortunately, even the basic idea of having a community controlled or government regulated institutional mechanism would be against the spirit of new economic policy which relies more on resource allocation by market forces rather than by the government. Likewise, removal of food subsidy may reduce the real income of poor farmers and their capacity to invest in agricultural modernisation. Similarly, the positive impact of the price decontrol measure would be uncertain because of either negative or poor possible response of small farms and backward regions to average price variation within and outside the country, in the absence of necessary technological and infrastructural support.

Moreover, export orientation of large farms and unrestricted growth of internal demand for foodgrains may push up the food prices and small and marginal farms who have to buy goodgrains at the lean season may experience deterioration in their real income position. Consequently, investment levels of small farmers would probably fall, thereby retarding growth of output on small farms. Even the scope of earning sufficient income and foreign exchange through commercialisation or export orientation of large farms is small because of lack of appropriate technological, infrastructure and organisational changes.

It is significant to note that rates of both public and private investment in agriculture have declined in recent years. While the farmers of developed regions are demanding further technological innovations to grow, in the agriculturally backward regions, the rates of private investment and technological change have been slow, partly due to inadequate public investment on infrastructure and partly due to lack of appropriate agrarian reform measures.

It is feared that the on-going process of structural reform may lead to further cut on public investment in agriculture, which in turn would reduce the rate of private investment, thereby slowing down the pace of agricultural growth. Moreover, it would be risky to depend on price decontrol and market reform for agricultural improvement. To quote John W. Mellor, 'Prices are indication of, not solution for the problems of agricultural production and employment...prices are extremely inefficient way to send signals (Mellor, 1986). According to Adrin Wood (1991), market determined domestic prices will not necessarily provide consistent or socially rational signals to farmers or consumers. It is often argued that left to market forces, favourable product prices would augment agricultural output and employment in general and marginal farmers may benefit from the increased employment opportunities thus created. In reality, however, supply response to price variation is very low in the case of good crops, particularly in infrastructurally under-developed regions and recently there has also been a decline in employment elasticity of agricultural output in many regions. Therefore, employment effect of price variation via output growth may not necessarily compensate for the loss of money of marginal farmers due to hike in food prices. In fact, small farmer's response to price incentives by way of increase in investment would depend on a variety of factors including their accessibility to land, capital, technology and physical and socio-economic infrastructures. In addition, the provision of agricultural trade liberalisation as laid down in the NEP and the WTO may lead to further hike in the prices of critical farm inputs, including technology which would affect the growth of agricultural output adversely. The view (Gulati and Bhide, 1993) that trade liberalisation would tend to promote favourable agricultural terms of trade and income of agriculturists through exports of agricultural commodities may prove to be an illusion, as the overall policy package may have unfavourable impact on the pace and patterns of agricultural growth. We need to identify and accept only those positive aspects of reform which promote growth and development through maximization of both demand and supply side effects within and outside the country. Besides, institutional efforts to generate and spread appropriate technology would hold the key to rapid agricultural growth (Swaminathan, 1983). Moreover, any major shift in the cropping patterns (however, desirable it might be for the nation), in favour of fruits, vegetables and other high value crops may not be attractive for the marginal and small farmers, unless foodgrains are made available to them at subsidised prices and various land, capital and market constraints are removed. The provision of appropriate pricing and subsidies would be essential atleast during the transition phase. Let Agri-business consortium help the small and marginal farmers in overcoming atleast the credit, technology, market and organisational constraints. In short, given the access to appropriate technology, credit, market, price/subsidy, guidance and non-farm work opportunities, small and marginal farmers may respond well to the needs of commercial agriculture on viable basis. However, in order to sustain on long-term basis, small and marginal farms must plan to reduce their family sizes and diversify their activities and expand non-farm sources of income.

REFERENCES

Gulati, Ashok and S. Bhide, 'New Economic Policy : Likely Implications for Agriculture, Paper presented at the annual AEAI Conference held at IARI, Feb., 1993 (mimeo).

Haque, T. (1988). 'Role of Technological Change in Improving the Economic Conditions of Marginal Farmers and Landless Labourers in India', Paper presented at the Seminar organised by IASSI, New Delhi, Nov., 1-3, 1988.

ICAR, Annual Report on National Demonstrations, New Delhi, 1986-87.

Kaneda, Hiromitsu, 'Structural Change and Policy Response in Japanese Agriculture after the Land Reforms', Economic Development and Cultural Change, Vol. 28(3), 1980.

Mellor, J.W. (1986). 'Agriculture on the Road to Industrialisation', IFPRI, Washington, D.C. Reprint No. 83, 1986.

Sharma, V.K. and T. Haque (1986), 'Impact of Fertiliser Subsidy on Agricultural Growth and Equity', IASRI, 1986.

Swaminathan, M.S. (1983). 'Science and the Conquest of Hunger', Concept Publishing Company, New Delhi.

Wood, Adrin (1991). 'China's Economic System with some suggestions for further reform, London School of Economics, C.P. No. 12, p. 32.

4

IMPACT OF FARM SIZE ON PRODUCTIVITY AND EMPLOYMENT

Probably no subject has been so much researched and debated by agricultural economists as the relationship between farm size and land productivity. Sen (1962) was one of the first to argue that small farms were relatively more productive, as there was an inverse relationship between size of farm and land productivity. Following Sen, a number of other economists including Mazumdar (1963), Hanumantha Rao (1966), Bhagawati and Chakravarthy (1969), Lau and Yotoplous (1971), Bhattacharya and Saini (1972), Bharadwaj (1974) and Hopper (1975) also observed that in Indian context, small farms were relatively more efficient than large farms. According to these economists, the factors accounting for higher productivity on small farms included (i) higher fertility (Sen, Rao, Bhagwati and Chakravarthy); (ii) higher level of managerial efficiency (Rao); (iii) relatively more intensive use of land, labour and other non-capital inputs (Sen, Mazumdar, Rao and Bharwaj); (iv) higher impact of indivisible factors (Rao); and (v) relatively high preference for leisure by big farmers (Rao).

However, in recent years, studies by Deolalikar (1981) and Parthasarathy (1987) showed that in green revolution regions, productivity had no significant relationship to farm size as large farms were found to invest more in modern inputs including HYV seeds, chemical fertilizers, pesticides and irrigation as compared to small farms. It is said that as a result of technological change, there has been a reduction in the importance of labour input and increase in the role of new inputs which are capital intensive in nature and less readily available to small farmers which explain why the relationship between farm size and land productivity is being either weakened or reversed.

In a more comprehensive study, Chattapodhaya and Rudra (1976) observed that inverse relationship held in many, but not all parts of the country, that it did not hold at all times and in all ranges of the farm size and that it did not hold for individual crops as there was no tendency for yields per acre to decline with farm size. According to Ghosh (1976) however, the inverse relationship between total output and farm size was still found to exist. More recently, a study by Bal, Jain and Singh (1991) revealed that in the agriculturally developed state of Punjab, marginal and small farms continued to have an edge over medium and large farms in terms of gross returns per unit of area, although the average net return (after deducting the cash expenses from gross returns) did not differ significantly as between various size groups of farms.

Thus, in view of the conflicting results of various studies mentioned above, it was considered appropriate to re-examine various issues relating to size-productivity relationship in greater details. Specifically, the following issues required more careful investigation :

(i) Cateris Paribus, does size-prôductivity relationship differ between technologically developed and underdeveloped regions?

(ii) Do we observe the size-productivity relationship to differ due to various productivity concepts including :

(a) Size of farm and yields of individual crops;
(b) Size of farm and gross/net returns per unit of area;
(c) Size of farm and labour productivity;
(d) Size of farm and overall factor-productivity;

(iii) What are the possible factors that influence size-productivity relationship? Do the differences in land quality, land tenure, management, access to modern inputs, technology, intensity of factor use, factor-product prices and the differential socio-economic goals and incentives of two sets of farms matter most or size by itself has any role to play?

(iv) Does size of farm influence occupational mobility and growth of non-farm income?

(v) Does size of farm influence the use of labour and patterns of employment in agriculture?

(vi) Does size-productivity relationship throw any light on the question of land reforms and desirable pattern of landholdings?

Answers to the above questions will largely help explode the myth regarding the varying impact of farm size on productivity, income and employment.

Does Size-Productivity Relationship differ between Technologically Developed and Under-Developed Regions?

In order to test the hypothesis that new technology weakens the traditional inverse relationship between farm size and productivity, an attempt was made here to find out the nature of size-productivity relationship in both technologically developed and underdeveloped villages situated under homogeneous as well as heterogeneous agro-climatic conditions. Both tabular and regression analysis techniques were used for this purpose. The following two equations were estimated to examine the role of size of farm in determining gross land productivity and that of fertilizer and human labour in influencing net returns per unit of net cultivated area.

(i) $Y_a = S_1 \beta_1 e^u$

(ii) $Y_n = F_1 \beta_1 N_2 \beta_2 e^u$

Where Y_a = Gross returns per unit of net cultivated area (ha.)

Y_n = Net returns per unit of net cultivated area (ha.)

S = Size of farm

F = Value of Fertilizer use per hectare

N = Value of Human labour per hectare

Table 4.1 shows the yield of rice (which was found being cultivated in all the selected villages, excepting those of Rajasthan) by various size groups of farms. It could be seen from the table that in both developed and under-developed villages of Andhra Pradesh, small and medium farms had relatively higher levels of yields. But in the case of Bihar villages, there was a mixed trend with negative size-productivity relationship in Miranand and Mathurapur villages of Bhagalpur and positive relationship between rice yield and size of farm in both Simra and Niyamatchak villages of Patna. In Haryana, the technologically under-developed villages of Bhiwani showed more or less similar trend as that of Andhra Pradesh, small and medium farms having an edge over both marginal and large farms. But in the developed district of Karnal, medium and large farms had relatively higher yields of rice as compared to both marginal and small farms. Thus, there was no uniform pattern of relationship between farm size and rice yields in various regions of the country. Still certain trends become evident. First, marginal as well as large farms produce less rice per unit of cropped area than small and medium farms, although the reasons for low rice yields on marginal farms could be different than those of large farms. While marginal farms suffer on account of capital and other input constraints, large farms may lack necessary managerial ability and incentives to produce more. Secondly, in most of the technologically developed areas, there is an emerging trend of medium size farms above 5 acres to produce more rice per unit of cropped area than all other categories of farms. This does invariably suggest that traditional inverse relationship between size of farm and crop productivity tends to weaken in most of the technologically developed villages studied. But the very fact that large farms produce less per unit of cropped area than medium farms in technologically developed villages and as compared to both small and medium farms in all other villages demonstrate the necessity of an appropriate size of farm, not unmanageably too large.

However, the size-productivity relationship as measured in terms of net returns per unit of net cultivated area shows a slightly different picture. It could be seen from Table 2.22 that in the case of Andhra Pradesh, small and medium farms in Anantapur villages and medium and large farms in Vatlure and small and large farms in Ponangi villages of West Godavari had comparatively higher levels of net returns per hectare of net cultivated area (from crop farming). In Bihar, there is almost a uniform trend with small and medium farms having comparatively higher net returns than both marginal and large farms in all the villages studied. In Haryana, there was a mixed trend. In the technologically developed district of Karnal, medium farms had an edge over all other categories of farms, but in the relatively underdeveloped district of Bhiwani, medium farms in Chandabass-I and marginal farms in Chandabass-II had received higher net returns per unit of net cultivated area. In Rajasthan, however, net returns were found to be relatively on large farms in the developed villages of Kyd and Netewala and on small farms in the under-developed village of Amarsar. It could be further seen from Table 2.21 that there is large scale inter-farm variation in net returns from crop production per unit of net cultivated area. In the case of Andhra Pradesh, co-efficients of inter-farm variations in the net returns ranged between 14.3 per cent in West Godavari to 30.3 per cent in Anantapur. But in the states of Bihar, Haryana and Rajasthan, the relatively developed districts of Patna, Karnal and Sriganganagar were found to have relatively higher co-efficients of inter-farm variation in net returns than those of the relatively underdeveloped districts of

Bhagalpur and Bhiwani. It may be further seen from Table 4.2 that elasticity co-efficients of size of holding as a determinant of farm productivity also indicate more or less above type pattern. In six out of fifteen villages, there was an inverse relationship between farm size and gross returns per hectare, while positive relationship between farm size and gross returns per hectare could be observed in the remaining eight villages studied. Besides, although small farms were using relatively less quantum of chemical fertilizers, marginal productivity/ elasticity co-efficients of this input on small farms were found to be relatively greater than those of large farms (Table 4.2).

TABLE 4.1 : Inter Size Class Variation in the Yields of Rice in Selected Villages

State	District	Village	Size Group	Yield of rice (q/ha.)
(1)	(2)	(3)	(4)	(5)
Andhra Pradesh	Anantapur	Chukkalur	Marginal	9.0
			Small	11.5
			Medium	9.2
			Large	9.0
		Chennampally	Marginal	8.2
			Small	9.5
			Medium	10.7
			Large	9.0
	West-Godavari	Ponangi	Marginal	40.0
			Small	41.4
			Medium	41.2
			Large	39.1
		Vatlure	Marginal	30.5
			Small	31.6
			Medium	32.0
			Large	29.2
Bihar	Bhagalpur	Miranand	Marginal	14.2
			Small	13.3
			Medium	13.0
			Large	12.8
		Mathurapur	Marginal	12.5
			Small	10.0
			Medium	9.7
			Large	9.7
	Patna	Simra	Marginal	10.5
			Small	12.6
			Medium	14.1
			Large	14.0

(Contd.)

TABLE 4.1 : (*Contd.*)

(1)	(2)	(3)	(4)	(5)
		Niyamat-chak	Marginal	10.0
			Small	12.1
			Medium	12.5
			Large	12.9
Haryana	Bhiwani	Chandabass-I	Marginal	14.5
			Small	15.1
			Medium	16.5
			Large	14.6
		Chandabass-II	Marginal	18.2
			Small	21.0
			Medium	20.7
			Large	18.9
	Karnal	Gagsina	Marginal	21.5
			Small	23.0
			Medium	25.8
			Large	25.2
		Munak	Marginal	21.8
			Small	22.0
			Medium	23.9
			Large	22.4

TABLE 4.2 : Elasticity Co-Efficients of Size of Land in Influencing Gross Returns per Hectare of Net Cultivated Area

State	District	Village	β	SE	R^2
Andhra Pradesh	Anantapur	Chukkalur	–0.13*	0.61	0.11
		Chennampally	–0.34	1.24	0.14
	West Godavari	Ponangi	0.15	0.35	0.24
		Vatlure	0.12**	0.01	0.21
Bihar	Bhagalpur	Miranand	–0.35**	0.09	0.33
		Mathurapur	–0.11	1.23	0.08
	Patna	Simra	0.08	0.04	0.12
		Niyamatchak	0.01	0.09	0.04
Haryana	Bhiwani	Chandabass-I	0.04*	0.01	0.13
		Chandabass-II	–0.12*	0.05	0.15
	Karnal	Gagsina	–0.02	1.14	0.08
		Munak	0.03	0.02	0.13
Rajasthan	Bikaner	Kyd.	0.15**	0.03	0.22
	Sriganga-nagar	Netewala	0.08	0.06	0.11
		Amarsar	0.02	1.03	0.06

* Statistically significant at 10 per cent level.

** Statistically significant at 1 per cent level.

But the very fact that some of the technologically developed villages showed strong inverse relationship between farm size and productivity, proves beyond doubt that under favourable environment, small and marginal farms adopt new technology and positive interaction between technology and high intensity of land and labour use result in higher yields on such farms. This finding clears the doubts raised by Parthasarathy and Deolalikar and proximates more to the findings of Bhalla (1979) and Ghonemy (1990) that given the access to favourable technology and infrastructure, small farms would respond well. The proportion of area under high yielding varieties was not much different as between various size groups of farms. The per hectare use of chemical fertilizers was relatively low on small and marginal farms in most of the villages studied. But in the three specific villages where there was significant inverse relationship between farm size and productivity, the inter-size group differences in the use of chemical fertilizers was negligible and the use of local manures was even higher on small and marginal farms. Also the proportion of area irrigated was not much different and small and marginal farms had access to institutional credit facilities. Thus, HYV technology per se cannot be said to weaken the inverse relationship between farm size and productivity. In places where small and marginal farms had adequate access to new technical inputs including HYV seeds, irrigation, chemical fertilizers and above all institutional credit facilities which enabled them to invest in modern inputs, inverse size-productivity relationship continued to exist. However, in areas where institutional credit facilities are inadequate and input delivery system is weak, large farms by virtue of their greater ability to invest in modern inputs seem to reap relatively higher productivity.

Size-Productivity Relationship under Various Productivity Concepts

It is often argued that size-productivity relationship would vary due to various productivity concepts. While yields/returns per unit of land may be higher on small farms, productivity per unit of labour is expected to be higher on large farms. Besides, inter-farm differences in net returns are expected to be lower than those of physical yields or gross returns per unit of area because of low marginal return on high level of input use on large farms. Therefore, an effort was made here to examine whether size-productivity relationship differs under various productivity concepts. As we have seen from Table 4.1 that yields of rice were generally higher on small and medium farms as compared to both marginal and large farms in majority of the villages studied. But average as well as marginal returns from labour input were invariably lower on small and marginal farms in both developed and under-developed villages. While there was an inverse relationship between farm size and labour use per unit of area, marginal productivity of labour increased with increase in the size of holding. Infact, large farms rely more on non-labour inputs for productivity growth and, therefore, marginal product of labour is generally high due to low intensity of labour use and its positive interaction with other technological and infrastructural factors.

Furthermore, it could be seen from Table 2.20 that co-efficients of inter-farm variation in net returns were higher than those of gross returns in all the villages studied. Thus, whereas size-productivity relationship seems to vary depending on whether we use the concepts of land or labour productivity, there is no clear trend or pattern due to the use of various land productivity concepts. However, the relationship between farm size and net returns per rupee invested differed significantly due to the use of various cost concepts. If we assume that family labour does not have an opportunity cost, then computation of net returns excluding family labour shows that net returns were invariably higher on small farms in all the villages studied. This is because marginal and small farms had used relatively higher amount of family labour. Besides, even if we decide to include the cost of family labour, it is important to decide a *priori* whether the cost of family labour, would be valued at the prevailing market wage rate or below it as the different procedures followed would yield different results. Nevertheless, it is reasonable to argue that family labour has a cost and there is no reason why family labour should be excluded while calculating net farm returns.

Factors Influencing Size-Productivity Relationship

Any scientific analysis of factors influencing size-productivity relationship would necessarily involve identification of factors which either promote or retard productivity growth and with respect to which each size group of farms are reported to have some relative advantages and disadvantages. Infact, in any given agro-climaţic situation, it is these factors which determine the nature of size-productivity relationship. The earlier studies by Sen (1964) Bhagwati and Chakravarthy (1969) explained the inverse relationship between farm size and productivity mainly by saying that small farms had cultivated relatively better quality of land and that they used land and labour more intensively. While large farms are said to have greater ability to invest in modern inputs and also bear risks of various types of uncertainties including violent changes in climate (leading to drought, flood, hailstorm etc.), factor-product pricing, technological and occupational diversification, small farms are generally reported to be using their land and labour more intensively. While large farms depend mainly on hired labour, small farms use as far as possible only family labour which by nature has greater incentive to work intensively and with greater care for productivity growth. Besides, whereas large farms are said to suffer from backward bending supply curve of efforts, small farms are supposed to use their land and labour more efficiently for the sake of mere existence. Consequently one hundred small farms of one hectare each would produce more output than one hundred acre farm of a large farmer. In addition, the socio-economic goals of different size groups of farms are said to differ to a large extent, and therefore, there is built in rigidity in land use pattern and technology in many cases which could be altered only through appropriate, albeit massive organisational as well as infrastructural changes. Moreover, it is also argued that farm size by itself cannot influence productivity. It is the overall socio-economic status of a farmer which is important. A farmer's technological innovativeness and risk bearing attitude as well as ability can be possibly changed through appropriate changes in infrastructural and institutional facilities of irrigation, credit, marketing and extension. Moreover, Binswanger and Rosenwig (1981) are of the view that owned area is positively correlated with land productivity, while operated area is negatively associated with land productivity, thereby emphasising the role of tenancy and tenancy reforms in influencing size-productivity relationship. The results of our farm level survey which throw adequate light on some of these issues could be summarised as follows :

Land Quality and Size Productivity Relationship

An explanation often given for the traditional inverse relationship between farm size and productivity is that small and marginal farms cultivate better quality of land. According to Bhagwati and Chakravarthy, while selling away land, farmers try to give away the less fertile land and keep the best land with them. This is how small farmers tend to possess better quality land and large farmers who increase their size of holding through purchase have relatively low quality of land. In reality however, this argument looks very weak. In all our study villages, farmers of all categories had some plots of land which were relatively inferior in quality, although because of less number of plots and low size of holding, the proportion of good quality land was generally higher on small farms. Nevertheless, there were also cases where marginal farmers cultivated land of very poor quality and received low yields of crops. This is particularly true of the ceiling surplus land assigned to the landless or semi-landless. Besides, inherent low soil fertility of medium and large farmers seems to have been significantly remedied through irrigation and application of micro and macro nutrients. While small farms had relatively higher proportion of area under irrigation, large and medium farmers had applied more macro as well as micro nutrients per unit of area. This is how it is difficult to prove that inverse relationship between farm size and productivity exists due to relatively poor quality of land cultivated by large farmers.

Intensity of Land Use as a Factor Influencing Size-Productivity Relationship

Both macro level aggregate data and our survey data show that cropping intensity is relatively higher on smaller farms and therefore, other things remaining the same, productivity per unit of net cultivated area is also higher on small farm (Table 4.3). Besides, area under fallow increases with size of holding and hence also the productivity per unit of available land is higher on small farms. However, in relatively developed villages of Haryana, presumably under the impact of mechanisation, such gap in cropping intensity tends to be lower. Nevertheless, it goes beyond doubt to prove that differential intensity of land use is one of the important factors influencing size-productivity relationship.

Intensity of Labour use and Size-Productivity Relationship

It is often argued that smaller farms are able to raise more output per unit of area because they use labour more intensively than large farms which depend more on hired labour that has a monitoring cost and which also lacks incentive to work. It is true that even some small farms hire labour in peak seasons, but they greatly depend on surplus family labour. It could be seen from Table 4.4 that other things remaining the

TABLE 4.3 : Intensity of Cropping by Size Group of Farms in Selected Villages

State	District	Village	Size Group	Cropping Intensity (%)
(1)	(2)	(3)	(4)	(5)
Andhra Pradesh	Anantapur	Chukkalur	Marginal	1.11
			Small	1.13
			Medium	1.08
			Large	1.05
		Chennampally	Marginal	1.10
			Small	1.08
			Medium	1.03
			Large	1.03
	West-Godavari	Ponangi	Marginal	1.89
			Small	1.92
			Medium	1.72
			Large	1.68
		Vatlure	Marginal	1.65
			Small	1.59
			Medium	1.45
			Large	1.44
Bihar	Bhagalpur	Miranand	Marginal	1.62
			Small	1.36
			Medium	1.29
			Large	1.20
		Mathurapur	Marginal	1.25
			Small	1.42
			Medium	1.20
			Large	1.21
	Patna	Simra	Marginal	1.58
			Small	1.80
			Medium	1.69
			Large	1.52
		Niyamat-chak	Marginal	1.42
			Small	1.46
			Medium	1.36
			Large	1.38
Haryana	Bhiwani	Chandabass-I	Marginal	1.50
			Small	1.60
			Medium	1.36
			Large	1.38
		Chandabass-II	Marginal	1.59
			Small	1.65
			Medium	1.50
			Large	1.40

(*Contd.*)

TABLE 4.3 : (*Contd.*)

(1)	(2)	(3)	(4)	(5)
	Karnal	Gagsina	Marginal	1.79
			Small	1.85
			Medium	1.72
			Large	1.70
		Munak	Marginal	1.59
			Small	1.61
			Medium	1.58
			Large	1.52
Rajasthan	Bikaner	Kyd.	Marginal	1.60
			Small	1.62
			Medium	1.38
			Large	1.35
	Sriganga-nagar	Netewala	Marginal	1.91
			Small	1.86
			Medium	1.62
			Large	1.60
		Amarsar	Marginal	1.10
			Small	1.08
			Medium	1.08
			Large	1.05

TABLE 4.4 : Inter-Size Class Variation in Labour Use per Unit of Net Cultivated Area in Selected Village

State	District	Village	Size Group	Labour days per ha. of net cultivated area	
				Male	Female
(1)	(2)	(3)	(4)	(5)	(6)
Andhra Pradesh	Anantapur	Chukkalur	Marginal	37	78
			Small	51	65
			Medium	53	58
			Large	48	55
		Chennampally	Marginal	48	60
			Small	41	56
			Medium	47	48
			Large	44	54
	West-Godavari	Ponangi	Marginal	276	302
			Small	278	296
			Medium	232	251
			Large	242	224
		Vatlure	Marginal	201	259
			Small	220	244
			Medium	220	195
			Large	221	185

(*Contd.*)

TABLE 4.4 : (*Contd.*)

(1)	(2)	(3)	(4)	(5)	(6)
Bihar	Bhagalpur	Miranand	Marginal	132	88
			Small	122	82
			Medium	105	64
			Large	103	55
		Mathurapur	Marginal	76	79
			Small	100	107
			Medium	92	66
			Large	89	63
	Patna	Simra	Marginal	244	207
			Small	313	225
			Medium	274	199
			Large	248	165
		Niyamat-chak	Marginal	159	153
			Small	186	140
			Medium	132	84
			Large	118	80
Haryana	Bhiwani	Chandabass-I	Marginal	50	52
			Small	72	69
			Medium	90	57
			Large	110	67
		Chandabass-II	Marginal	72	78
			Small	68	70
			Medium	84	69
			Large	86	71
	Karnal	Gagsina	Marginal	67	65
			Small	81	78
			Medium	108	66
			Large	96	54
		Munak	Marginal	51	70
			Small	66	60
			Medium	94	60
			Large	87	47
Rajasthan	Bikaner	Kyd.	Marginal	59	82
			Small	53	57
			Medium	88	61
			Large	97	52
	Sriganga-nagar	Netewala	Marginal	82	89
			Small	101	96
			Medium	82	67
			Large	67	45
		Amarsar	Marginal	55	59
			Small	59	62
			Medium	57	53
			Large	52	48

same, use of human labour per unit of net cultivated area is invariably higher on smaller farms in both developed and under-developed regions. In the relatively developed state of Haryana, both small and large farms use comparatively less labour per unit of area than other developed districts of Patna and West Godavari, presumably under the impact of mechanisation. But the proportion of female labour use per hectare is greater on small farms everywhere including Haryana. Also, as we have seen in Table 2.21 average land productivity is higher on small farms only in some and not in all villages, thereby implying that positive effects of intensive labour use gets neutralised by the use of other technological factors that result in higher productivity. In some cases, medium and large farms operate more efficiently by displacing labour through mechanisation, while small farms tend to use the available family labour for maximum yields because of lack of adequate/ alternative employment opportunities. Besides, in most cases, it is due to the small farmer's bare necessity to earn more for subsistance through hard labour from a fixed small unit of land that makes all the difference. On the whole however, there is no denying the fact that small farms tend to employ more labour, (particularly the family labour) per unit of area, especially in places where alternative employment opportunities are not either adequate or acceptable to small farm family labour.

Does Ability to Invest or Bear Risk Matter?

It is often said that in the context of new production technology, it is only the large farmers who have the necessary ability to invest in modern inputs and take full advantage of the available technological potentials for raising farm productivity. Besides, it is said that due to higher size of land holding, only large farms can bear the risks of income loss due to crop failure which occurs almost at regular intervals in a cyclical manner in most of the rainfed or drought prone areas. Moreover, because of large size of holding, large farms are said to have potentials to minimise such risk through diversification and rational allocation of land between crop cultivation, horticulture, sericultrue, aquaculture, animal husbandry etc. In course of our inquiry, it was observed that medium and large farms had relatively higher levels of farm business income and savings, and therefore, greater ability to invest. But this did not always mean higher investment in modern inputs per unit of area, as small farms also were found to invest in irrigation water and chemical fertilizers as much as those of large farmers, particularly in places where infrastructural and institutional facilities of credit, irrigation, extension and markets were available. It is, however, true that whenever there is any crop failure due to flood, drought or pest attack, small and marginal farmers suffer most, particularly those who cannot migrate to safer zone for hiring out labour. However, it was also observed that under normal condition, small farmers tended to diversify their economic activities more for supplementing their income for subsistence and by resorting to animal husbandry, poultry, sericulture and even by taking up non-agricultural employment outside on temporary basis. Thus, while large farms have no doubt greater ability to invest and bear risks which help promote farm productivity, small farms also have the necessary will to invest in new technical inputs and bear risk, provided infrastructural and institutional facilities of credit, irrigation, extension and market and insurance are provided.

Tenancy and Size-Productivity Relationship

The Binswanger-Rosenwing (1981) hypothesis that owned area is positively correlated with land productivity, while operated area is negatively associated with land productivity, was tested by examining the role of tenancy in influencing size-productivity relationship. It was observed that although tenancy influenced size-productivity relationship in various ways, Biswanger-Rosenwig hypothesis could not be proved in many cases. Even when the size-productivity relationship was examined on the basis of owner ship holdings, the results were not significantly different than what we observed on the basis of operational holdings. This is not to suggest that tenancy does not affect farm productivity adversely. Infact, our macro level analysis shows that there is an inverse relationship between farm size and tenancy on the one hand and farm productivity on the other (Haque, 1992).

Does Size of Farm Influence Migration/Diversification?

It is true that, farming alone cannot sustain all the growing millions in Rural India and there is need for growth of income and employment outside of agriculture. According to Census of India, the percentage of rural workers engaged in non-farm activities increased only marginally from 18.9 in 1981 to 19.8 in 1991. While the proportion of male workers rose marginally from 20.5 in 1981 to 22.1 per cent in 1991, that of female workers declined from 13.6 per cent in 1981 to 13.1 per cent in 1991. Unfortunately, the intensity of labour use in agriculture has shown a declining trend in recent years in many states. Table 3.4 shows the changes in labour use per hectare over time for various crops.

Our micro level analysis (Table 4.5) however, shows that marginal farmers in general depend on employment outside their own farms for subsistence. While marginal farmers were found to migrate and take up non-agricultural occupation for subsistence, some large farmers also were found to migrate to non-farm occupation on both part time and full time basis. The main reason for migration of large farmers was the lack of adequate investment opportunities in agriculture and the consequent urge to earn income through investment in non-farm sector. In our study villages, small and medium farmers seem to be more attached to land and less migratory in character. Even for large farmers, the prospect of migration was non-existent in most of the villages studied. This could be due to both backward bending supply curve of effort on the part of large farmers and risk involved in diversification, which requires more investigation.

Does Size Productivity Relationship Throw any Light on the Question of Land Reforms and Desirable Pattern of Land Holdings?

An economic case for land reform has generally been made on the basis of size-productivity relationship. Since our survey results indicate that inverse relationship between farm size and productivity continues in most of the regions studied, we may argue that there is still a valid case for further redistribution of land on grounds of efficiency. Besides, as small and medium farms were found to raise higher yields per unit of area than both large and marginal farms, it may be desirable to take away land from very large holdings for their distribution among marginal farms so that marginal farms could be converted into small, albeit efficient ones. But given the rising pressure of population on land, it is likely that most of our farms would tend to become marginalised in the near future and without sufficient technological breakthrough in agriculture and development of non-farm employment, land reforms alone may fail to yield the desired result. However, both marginal and large farms were found to have greater potentials for migration outside agriculture than small and medium farms who were more attached to land. While marginal farmers tend to migrate because of non-viability of their farms and economic compulsion to earn additional income by hiring out labour in neighbouring towns/factories, the large farmers if forced (through radical land reform) to migrate would

TABLE 4.5 : Proportion of Agricultural Migrants During the Past Five years to other Non-Agricultural Occupation in Various Size Class of Farms

State	District	Village	% Migrants							
			Marginal		Small		Medium		Large	
			Full time	Part time	Full time	Part time	Full time	Part time	Full time	Part time
Andhra Pradesh	Anantapur	Chukkalore	3.2	9.9	1.3	6.2	2.2	1.0	2.0	2.5
		Chennampally	2.1	10.1	1.3	7.5	1.2	—	12.0	—
	West-Godavari	Ponangi	—	5.5	—	—	—	3.7	5.0	15.2
		Vatlure	—	5.0	—	—	—	3.2	1.2	10.0
Bihar	Bhagalpur	Miranand	—	—	—	—	—	—	—	—
		Mathurapur	—	—	—	—	—	—	—	—
	Patna	Simra	2.4	15.1	2.3	—	—	—	—	—
		Niyamatchak	—	12.0	2.0	—	—	—	—	—
Haryana	Bhiwani	Chandabass-I	—	38.0	—	2.0	—	—	—	—
		Chandabass-II	—	30.1	—	—	—	—	—	—
	Karnal	Gagsina	—	2.2	—	—	—	—	—	—
		Munak	—	3.1	—	—	—	—	—	—
Rajasthan	Bikaner	Kyd	5.2	10.1	5.0	7.8	—	—	5.0	1.0
	Sriganganagar	Netewala	—	—	3.2	6.2	—	—	—	2.0
		Amarsar	10.5	25.0	5.1	—	—	—	—	1.0

depend for income creation on their entrepreneurial abilities rather than landed asset. This kind of logic further strengthens the economic case for land reform.

REFERENCES

Bal, H.S., K.K. Jain and Bant Singh. An Economic Analysis of Farm Business in Punjab, *Agricultural Situation in India*, July 1919, pp. 191-201.

Bhagwati, J.N. and Chakravarthy, S. (1969). 'Contributions to Indian Economic Analysis—A Survey' *American Economic Review Supplement*, September, 1969.

Bhalla, G.S. in Berry and Cline Edited (1979). 'Agrarian Structure and Productivity in Developing Countries, John Hopkins University Press, Baltimore, U.S.A.

Bhardwaj, Krishna 'Notes on Farm Size and Productivity' E.P.W., *Review of Agriculture*, March 30, 1974.

Bhattacharya, N. and G.R. Saini (1972). 'Farm Size and Productivity—A Fresh Look', E.P.W., June 24, 1972.

Binswanger, Hans and M.R. Rosenweig (1981). 'Contractual Arrangements, Employment and Wages in Rural Labour Markets : A Critical Review, New York, ADC.

Chattopadhya, M. and Ashok Rudra (1976). 'Size-Productivity Revisited', *Economic and Political Weekly*, Review of Agriculture, September, pp. A104-106, 1976.

Deolalikar, Anil B. (1981). 'The Inverse Relationship between Productivity and Farm Size—A test using Regional Data from India', *American Journal of Agril. Economic*, Vol. 63, No. 2 May 1981, pp. 275-79.

Ghonemy, M. Raid El. (1990). 'The Political Economy of Rural Poverty—The Case for Land Reform, Routledge, London & New Your, 1990.

Ghosh, A.K. (1979). 'Farm Size and Land Productivity in Indian Agriculture, *A Reappraisal' Journal of Development Studies*, October, 1979, Vol. xvi, No. 1.

Haque T. and Earl, O. Heady (1982). 'Problem of Resource Allocation in Indian Agriculture, Center for Agriculture and Rural Development, Iowa State University, U.S. (mimeo), 1982.

Haque T. (1992). 'Economics of Agriculture in Backward Regions, with special reference to viability and sustainability of Small and Marginal Farms', NIRD, Hyderabad (mimeo).

Hopper, David (1975). 'Allocation Efficiency in Traditional Indian Agriculture', *Journal of Farm Economics*, 1975.

Lau, L.J. and P.A. Yutopolous (1971). 'A Test for Relative Efficiency and Application to Indian Agriculture', *American Economic Review*, 1971.

Parthasarathy, G. (1987). 'Changes in the Incidence of Rural Poverty and Recent Trends in Some Aspects of Agrarian Economy', *Indian Journal of Agricultural Economics*, Vol. 42(1),1987.

Rao, C.H. Hanumantha (1966). 'An Alternative Explanations of the Inverse Relationship between Farm Size and Output per acre in India', *Indian Economic Review*, Vol. I, 1966.

Sen, A.K. (1962). 'An Aspect of Indian Agriculture' *Economic Weekly*, Feb. 1962.

5

IMPACT OF LAND TENURE ON FARM PRODUCTIVITY

The question whether agricultural productivity, income and employment get influenced by various forms of land tenure and farm size, has been a subject of academic discussion for a very long time. However, the results of past studies have thrown up some controversial issues which need to be carefully re-examined for arriving at meaningful conclusions. Besides, most of the past studies seem to have analysed the relationship between land tenure and productivity mainly with reference to yield differences of share cropped land and owner operated land, over looking various other forms of land tenure that exist due to any particular socio-economic environment and interlinkages among land, labour, credit and product markets. The present chapter is devoted to the re-examination of such issues on the basis of latest data collected from various agro-climatic regions of the country. This is intended to establish the true, albeit complex nature of relationship between land tenure and farm size on the one hand and agricultural productivity, income and employment on the other, which in turn would allow us to examine the relevance of land reforms to agriculture and rural development.

Alternative Forms of Land Tenure and Farm Efficiency

Although share cropping as an economic institution was criticised by all important classical economists, it was not clearly recognised until the time of Alfred Marshall that the share tenant would be an inadequate as well as inefficient cultivator. However, the logical basis of such inefficiency arguments has been under attack by several economists like Cheung (1969), Hsiao (1875) and Newbery (1974; 1977) saying that the argument is misleading since both landlord and tenant voluntarily contract with each other. Cheung (1968) observed that the implied resource allocation under private property rights is the same whether the landowner cultivates the land himself, hires farm lands to do the tilling, leases his holdings on a fixed rent basis or shares the actual yield with his tenant. The Chenguian model has been summarised by Bardhan (1980) as follows. Suppose that the landlord has one unit of supervisory labour at his disposal and would like to maximise his rental income from land. The control variables are the number of parcels he would like to make out of his land holdings (m), the share rental rate (r) and the labour intensity on the share cropped land (n^S). The production function of the share cropping tenant (f^S) is a linear homogenous function of the acreage of land rented in (k^S) and the quantum of labour devoted to it (n^S). The rental income of the landlord can be expressed as $R = mrf^S$. The main constraint of the landlord is that the share cropper's earnings cannot be lower than his alternative earnings as a wage labourer ($wn^S - (i - r)\ f^S$). Now the optimising conditions of the share cropper would be :

$$\text{to choose } (m, n^S\ r) \qquad \text{...(iii)}$$
$$\text{to maximise } R = mr\ f^f\ (k^s, n^s) \qquad \text{...(iv)}$$
$$\text{subject to } wn^s - (i - r)\ f^S.$$

If the labour intensity of the share cropped land is less than that under self-cultivation through hired labour, the landlord would prefer to earn higher income either by way of self-cultivation through hired labour or by fixed rental tenancy. Conversely, if the landlord insists on a higher labour intensity by share cropping tenant than what would be existing under wage cultivation, then there would not be any share cropper available. Therefore, the labour intensity of the share cropped land should be such that the marginal product of labour is equal to the wage level and the rent per unit of land is equal to its marginal product : Using general equilibrium approach, Bardhan and Srinivasan (1970) however, allowed both landlord and tenant to influence the share rental, while retaining the competitive labour market assumption of Marshall and Cheung. The optimising problem of the Bardhan-Srinivasan share cropper can be written as :

$$\text{to choose } (n_1{}^s, n_2{}^s, k^s)$$
$$\text{to choose } u^s\ (y^s, 1 - n_1{}^s - n_2{}^s) \qquad \text{...(iv)}$$
$$\text{subject to } y^s = (1 - r)\ f^s\ (k^s, n_1{}^s) + wn_2{}^s \qquad \text{...(v)}$$

Here n_1^s and n_2^s refer to the labour devoted to leased in land and the labour devoted to wage employment respectively. The tenant's utility function (u^s) is measured in terms of income (y^s) and leisure ($1 - n^{1s} - n_2^s$). The solution to the optimising problem gives the optimal values of n_1^s, n_2^s and k^s. It is further assumed that the demand for land k^s is an inverse function of the rental rate r. Considering the supply side, the landlord has the option of cultivating his own land with hired labour or leasing out land to the share cropper. The landlord like the tenant, would maximise his utility in terms of income and leisure which presumably determine the supply function of land for share cropping. The first order conditions for maximising (iv) subject to (v) provides us (after some cancellation) with :

$$(1 - r)\, f_2^s = w \qquad \text{...(vi)}$$
$$(1 - r)\, f_1^s = 0 \qquad \text{...(vii)}$$

Thus Bardhan and Srinivasan model also retained the zero marginal productivity of land condition which was criticised by Newbery (1974) and others as being implausible in land scarce economies. According to Newbery (1974), in most production functions including the popular Cobb-Douglas production function, equilibrium does not simply exist. Following Newbery (1974), Reid (1975) provided a generalisation of the Cheung model and concluded that the choice of tenure arrangement becomes inconsequential in the absence of uncertainty and transaction costs. However, the fact whether share cropping could be equally efficient would depend on whether the labour intensity stipulation emphasised by Cheung can be enforced, as the landlord might find it more convenient and economical to employ hired labour rather than to supervise the share cropper for enforcing the desired labour intensity.

A carefull examination of various issues raised by Marshallian as well as Cheungian Schools would reveal that their formulations represent over-simplification of the land tenure system in India where efficiency question of alternative land tenure arrangements cannot be analysed out of the context of various intersectoral linkages and interactions. Broadly speaking, there are ten major types of land relations in India which can be listed as follows :

(i) Owner cultivation where land is cultivated by land owner, using mainly family labour. This is denoted as T_1.

(ii) Owner cultivation where land is cultivated by the land owner, using solely or mainly hired labour. This is denoted as T_2.

(iii) Partly owner and partly tenant cultivation where marginal and small farmers own some land and also lease in land from absentee landlords for augmenting their income and employment levels. This is denoted as T_3.

(iv) Partly owner and partly tenant cultivation—where marginal and small farmers own some land and also lease in land from resident large farmers. The difference between this and the above T_3 type land tenure is that unlike absentee landlords, resident large farmers may monitor and supervise the cultivation work of tenants. This is denoted as T_4.

(v) Partly owner and partly tenant cultivation—where large and medium farmers own sufficient land and yet lease in land from absentee landlords for further strengthening of their income and power positions. This is denoted as T_5.

(vi) Partly owner and partly tenant cultivation—where large and medium farmers own sufficient land and yet lease in land from resident marginal and small farmers who are either induced or coerced to take up wage paid employment within or even outside the farm sector. This is denoted as T_6.

(vii) Partly owner and partly tenant cultivation where large as well as small and marginal farmers lease in land which are under usufructuary mortgage with professional money lenders or traders who cannot cultivate land themselves. In tribal areas, this is observed to be one of the major responsible factors for alienation of tribal land by the non-tribal money lenders and traders. This is denoted as T_7.

(viii) Partly owner and partly tenant cultivation—where small and marginal farmers lease in land from the disabled, widows, minors and army personnel. This is denoted as T_8.

(ix) Pure tenant cultivation—where agricultural labourers who do not own any land, lease in land from all classes of categories of landowners for augmenting their income and employment levels. This is denoted as T_9.

(x) Pure tenant cultivation where landless labourers lease in land for cultivation mainly from those land owners who provide wage employment or consumption credit to them. This is denoted as T_{10}.

The terms of lease also vary widely from region to region—depending not only on land relations, but also on the levels of technological development, infrasturctural and institutional facilities as well as the nature and extent of risks and uncertainties associated with farming in any given agro-climatic situation. These include : (i) share cropping where the landowner leases out land in return for a contractually stipulated portion of the output (the contract being either written or oral and may be legal or even illegal). This is denoted as T_{11}. (ii) fixed rent where land owner lease out land for a pre-determined rental charge per unit of area (in either cash or kind). This is denoted as T_{12}. (iii) Kisheni where land owners employ farm servants to cultivate land for a share or a fixed quantum of produce. This is denoted as T_{13}. (iv) land under usufructuary mortgage where the land is mortgaged to the money lender who may be a farmer or even a professional money lender who in turn leases out the land for cultivation either to the same land owner or some other cultivator for a pre-determined share or fixed quantum of produce or cash. This is denoted as T_{14}.

Infact, when we are discussing the relationship between land tenure on the one hand and agricultural efficiency and economic development on the other, all the above mentioned alternative forms of land tenure and terms of lease should enter into the model as explanatory variables. Besides, both in the supply and demand for leased land, it is not so much the question of monopolistic landlord versus competitive landless labourers which is important, but a variety of above mentioned relationships including inter-linkages between land, labour, credit and product markets which determine the nature and extent of lease.

Thus the traditional debate on efficiency of owner operated farms *vis-a-vis* tenant operated farms looks highly simplistic as well as unrealistic. In fact, alternative forms of land tenure as outlined above have different implications for productivity, employment and economic efficiency and therefore, an attempt has been made here to examine their relative merits and demerits in the context of resource use efficiency and economic development of the districts/villages under study. Theoretically speaking, the effects of the above mentioned alternative forms of land tenure and terms of lease may be analysed in the following manner.

Ceteris Paribus, the system of owner cultivation where land is cultivated by the land owner mainly with the help of family labour (*i.e.*, T_1) is supposed to be most efficient, provided the family labour does not have a high opportunity cost in terms of possible higher wage income outside the self-operated farms and that he has adequate access to modern technical inputs, credit, market etc. The system of owner cultivation where land is cultivated by the land owner solely with the help of hired labour (*i.e.*, T_2) would be relatively inefficient as he has to constantly supervise the work of hired labour which, therefore, has a transaction cost, unlike T_1 where family workers work for themselves. Besides, the owner does not work himself, but he merely supervises the work of others. Thus, to the extent he wastes his labour time and incurs a cost of supervision, T_2 category of farms would be inefficient as compared to T_1 category. The third category of farms involving partly owner and partly tenant cultivation where marginal and small farmers lease in land from absentee landlords (using mainly family labour) are supposed to possess a dual character *i.e.*, relatively higher intensity of use of labour as well as other inputs and consequently higher productivity on owned land and lower use of labour and other inputs and lower productivity on leased in land. This is particularly so because generally speaking, absentee landlords neither supervise the work nor contribute anything towards cash expenditure on modern inputs and yet claim 50 per cent of the total crop output. This also should act as a disincentive for the producers to cultivate tenanted land optimally. The fourth type of farms, *i.e.*, T_4 which falls in the category of partly owner and partly tenant, are also basically family farms. But the above type dualistic character of high yields on own land and low yields on tenanted land is hardly observed because unlike the absentee landlords, resident large farmers effectively monitor the cultivation work of tenants. Also there is sharing of inputs by the landlords which induces or coerces the tenants to apply more inputs for obtaining the best yields. In any case, the efficiency levels of such farms would tend to be higher than that of T_3 and closer to T_2. In a situation, where large and medium farmers own sufficient land and yet lease in land from absentee landlords for further strengthening of their income and power positions (T_5), cultivation is done mainly with the help of hired labour and would have the tendency to follow the efficiency level of T_2. Also when large farmers lease in land from the absentee landlords, the form of lease tends to be in favour of fixed produce or cash and not so much for share cropping which is the worst form of tenancy. T_6 represents partly owner and partly tenant farms where large and medium farmers own sufficient land and yet lease in land from marginal and small farmers who are either induced or coerced to take up wage paid employment within or even outside the farm sector. This generally occurs in irrigation command areas under consolidation programmes or when marginal and small farmers find it more rewarding to lease out the land for fixed rent and take up wage paid employment, for maximising their income/utility function. However, in many cases such small and marginal land owners who lease out land to large farmers tend to loose their land rights over time, due to either inducement or coercion by the large farmers who adopt various legal and illegal methods for alienating them from land. T_7 is a mixed type of farms representing partly owner and partly tenant cultivation where both large and small farmers lease in land which are under usufructuary mortgage with professional money lenders or traders who can not cultivate land themselves. Although this is not necessarily an inefficient form of land tenure, the land relations is highly exploitative and often leads to alienation of the poor and the tribals from land. The partly owner and partly tenant cultivation where small and marginal land owners lease in land from the disabled, minors, widows and army personnel (which are permitted by law) tend to be as efficient or inefficient as T_3, but certainly less efficient as compared to T_4 where resident large farmers effectively monitor the cultivation work of tenants. The pure tenant cultivation where landless labourers lease in land from all classes and categories of land owners for augmenting their income and employment levels is generally considered to be inefficient as compared to all other forms of land tenure, although, the actual outcome would depend on whether the land owner is a resident large farmer or absentee landlords, whether he effectively supervises the work of tenants or not and also whether the land is under fixed produce or share cropping. The share cropping, albeit pure tenants or absentee land owners, would be generally inefficient, because of the absence of both ability and incentive on the part of tenants to invest in modern inputs. But if the land is leased out for a fixed produce or cash rent, then atleast incentive effect could be taken care of. T_{10} also represents a case of pure tenant farms where landless labourers lease in land for cultivation mainly from those who provide wage employment or credit to them. This is generally performed under an arrangement in which three-fourth of crop produce goes to the land owners and one-fourth of it is received by the Kisheni or share cropping pure tenants who are often known as farm servants. The efficiency of such farms depends on the level of effective monitoring and sharing of inputs by the landowners. Nevertheless, monitoring involves cost which could be eliminated by conferring ownership right on the tenants.

The efficiency question of alternative land tenure arrangements could be explained in several ways. First, one simple way would be to compare the differential gross productivity per unit of land under alternative land tenure arrangements, before deducting the rent or landlord's share of produce. Second, the productivity comparison should be made after deducting the rent or landlord's share of produce. Third, the intensity of factor use including land, labour, capital as well as new technical inputs and their marginal productivity under varying land tenure systems could be compared.

This could be done by using either tabular method or the regression analysis technique in the following manner.

$$O = £\, T_1^{\beta 1}\, T_2^{\beta 2}\, T_3^{\beta 3}\, T_4^{\beta 4}\, T_5^{\beta 5}\, T_6^{\beta 6}\, T_n^{\beta n}\, e^u$$

Where O is gross value of output of a given land tenure system and T_1 to T_n represent proportion of area under various forms of land tenure. The elasticity co-efficients *i.e.*, β_1's would indicate whether a percentage change in acreage under any land tenure arrangement would influence a percentage increase or decrease in the value of gross output.

However, for the purpose of this study, the above mentioned equation could not be estimated because of inadequate number of observations in certain categories of tenure in the selected villages. A cross section estimation would have been possible, which might not make much sense due to heterogeneous agro-climatic environment. Therefore, a larger number of sample in a homogenous agro-climatic/ agro-climatic environment would be necessary. Nevertheless, the results of tabular analysis are quite revealing. It was observed that proportion of leased in area varied widely from region to region and between various size groups of farms. It ranged from about 1.98 per cent in Gagsina, the relatively developed village of Karnal district to 52.94 per cent in Kyd which is also a developed village of Bikaner district. In the case of marginal farms, the proportion of leased in area varied between 0.79 in Netewala village of Sriganganagar and 71.22 per cent in Ponangi village of West Godavari District and in the case of small farms, it ranged from 0.11 per cent in Chandabass-II village of Bhiwani to 82.9 per cent in Miranand village of Bhagalpur. It was observed that marginal and small farms increased their operational size of holdings by leasing in from others in most of the places, excepting in Amarsar and Kyd villages of Sriganganagar and Bikaner where medium and large farms leased in land substantially (Haque, 1992).

Table 5.1 shows the productivity of crops by tenure groups. It could be seen from the table that in Anantapur district of Andhra Pradesh,

TABLE 5.1 : Productivity of Crops by Tenure Group (Rs./ha. as of 1990-91)

State	District	Village	T_1		T_2		T_3		T_4		T_5	
			Gross Returns	Net Returns	Gross Returns	Net Returns	Gross Returns	Net Returns	Gross Returns	Net Returns	Gross Returns	Net Returns
Andhra Pradesh	Anantapur	Chukkalur	3299	1853	3057	1802	3048	394	3095	308	2932	720
		Chenam-pally	2081	1028	1834	1022	2070	119	2074	115	1899	222
	West-Godavari	Ponangi	13198	6085	15319	8008	13063	236	13406	192	13348	2953
		Vatlure	11726	6298	13863	7325	11488	96	11629	65	11974	2914
Bihar	Bhagalpur	Miranand	6032	3349	6129	2687	5907	509	6018	56	6114	385
		Mathurapur	4814	2992	5168	1722	4714	861	4746	136	4728	216
	Patna	Simra	6448	4346	6302	3990	6266	1001	6299	837	5794	625
		Niyamat-chak	5007	2894	6123	3862	4868	599	4948	587	4880	583
Haryana	Bhiwani	Chanda-bass-I	3213	2234	3580	2432	3145	739	3204	684	3414	649
		Chanda-bass-II	4230	2453	4320	2478	4138	786	4165	761	4066	150
	Karnal	Gagsina	7016	4609	8326	5013	6260	769	6916	737	6455	120
		Munak	5226	3415	6006	4057	5101	748	5152	708	4689	190
Rajasthan	Bikaner	Kyd	4950	3112	6813	4384	4866	600	4908	587	5638	1949
	Sriganga-nagar	Netewala	6313	4653	8814	6085	6186	1434	6225	1387	6698	2884
		Amarsar	2923	1691	3289	1759	2732	130	2825	42	2961	1183

(*Contd.*)

TABLE 5.1 : (*Contd.*)

State	District	Village	T_6		T_7		T_8		T_9		T_{10}	
			Gross Returns	Net Returns	Gross Returns	Net Returns	Gross Returns	Net Returns	Gross Returns	Net Returns	Gross Returns	Net Returns
Andhra Pradesh	Anantapur	Chukkalur	3133	1624	—	—	3006	1174	2954	566	3259	290
		Chenam-pally	1919	784	—	—	1914	541	1860	2	2023	16
	West-Godavari	Ponangi	12161	2546	—	—	—	—	11784	234	12570	527
		Vatlure	10739	2857	—	—	—	—	10425	0	11112	145
Bihar	Bhagalpur	Miranand	5500	1337	—	—	—	—	5298	601	5706	168
		Mathurapur	4430	232	4431	250	4709	124	4582	260	4596	44
	Patna	Simra	5933	659	—	—	—	—	5488	430	6123	813
		Niyamat-chak	4933	640	—	—	—	—	4456	348	4485	130
Haryana	Bhiwani	Chanda-bass-I	3188	601	—	—	3173	433	3038	330	3148	607
		Chanda-bass-II	3912	120	—	—	3974	599	3961	542	3968	456
	Karnal	Gagsina	6458	32	—	—	6146	922	6126	834	6144	836
		Munak	4730	130	—	—	5101	687	5035	606	5102	621
Rajasthan	Bikaner	Kyd	4936	1656	—	—	4851	598	4535	455	4728	532
	Sriganga-nagar	Netewala	5757	1699	—	—	5579	1140	5235	942	5357	1019
		Amarsar	2932	953	—	—	2662	194	2655	145	2738	175

per hectare gross as well as net returns of family farms (T_1) were higher than those of capitalist farms (T_2), presumably due to low level of technological development in the region and greater importance of human labour than other modern inputs. But in the technologically developed villages of West Godavari, both gross and net returns were higher on capitalist farms (T_2) having higher levels of gross returns per hectare than that of family farms (T_1). But the per hectare net returns were higher on family farms, indicating thereby relatively lower factor productivity on farms that depend mainly on hired labour (T_2). In both Haryana and Rajasthan, family farms (T_1) produced less per hectare of land than capitalist farms (T_2). Thus, on the whole, there was an emerging trend of capitalist farms producing more per unit of cropped area than family farms, excepting in the relatively very dry, albeit backward regions of Anantapur where capitalistic development in agriculture has yet to take place due to non-availability of appropriate farm technology. One possible explanation for such trend would be the relatively lower access of marginal family farms to technology, institutional credit and other inputs.

T_3 which represents marginal and small farms leasing in land from absentee land-owners, showed invariably lower gross returns per hectare than both T_1 (family farms) and T_2 (capitalist farms) in all the villages studied. Also the net returns per hectare on T_3 were very low because a substantial part of the gross produce was paid as land owner's share or rent. T_4 *i.e.*, marginal and small farms leasing in land from resident big land-owners, realised relatively more gross returns than T_3 group of farms, but net returns per hectare on T_4 groups of farms were found to be comparatively lower, presumably due to higher levels of input use and lower levels of input productivity. T_5 and T_6 show the operation of large and medium farms in lease market which is often termed as 'commercial' or 'reverse' tenancy. In the case of T_5 where large and medium farms leased in from absentee land owners, the per hectare gross returns were generally lower than those of T_1, T_2, T_3 and T_4 excepting in Vatlure villages of West Godavari and Kyd, Netewala and Amarsar villages of Rajasthan where T_5 group of farms realised relatively less gross returns than T_2 and more per hectare gross returns than all other categories of land tenure. Also the net returns per hectare of T_5 farms showed a mixed picture. It was found to be higher than those of T_3 and T_4 in the villages of Andhra Pradesh and Rajasthan. But in the villages of Bihar and Haryana, no such clear trend could be observed. While in Bihar, the net returns per hectare of T_5 was higher than that of T_4 and less than that of T_3, in Haryana, the net returns of T_5 were lower than those of both T_4 and T_3. The inter-regional differences in the net returns of T_5 could be explained by the varying terms of lease prevailing in different regions such as crop sharing, cost-sharing, fixed produce, fixed cash rent etc. For instance, large farms in both Andhra Pradesh and Rajasthan lease in land mainly against fixed produce or fixed cash, whereas marginal and small farms in general leased in land on crop sharing basis. The per hectare gross returns of T_6 showed a mixed trend. It was higher than T_5 in the villages of Anantapur, Patna and Karnal and less than T_5 in the villages of West Godavari, Bhiwani, Bikaner and Sriganganagar. But net returns of T_6 were lower than those of T_5 in eleven out of fifteen villages studied. T_7 group of farms could be seen only in the relatively under developed village of Bhagalpur and therefore, its productivity levels could not be compared with those of other regions.

T_8 represents leasing in land by small and marginal farms from disabled, army personnel, widows and minors. Although such cases of land tenure were few in number of most of the villages, the per hectare gross and net returns were compared here. It could be seen from the table that per hectare gross returns of this group of farms were generally lower than those of other groups excepting that in Anantapur where this group of farms realised relatively high gross as well as net returns than those of T_5.

T_9 and T_{10} represented landless agricultural labour class who leased in land from all classes of farms and under imperfect/interlinked factor market. The gross returns per hectare of T_9 group of farms were comparatively lower than those of T_{10}, but per hectare net returns showed a mixed trend and varied widely from region to region, depending on various terms of lease.

It is thus evident from the analysis of available data that farm productivity varied between land tenure groups. In technologically developed regions, capitalist farms (T_2) were found to have relatively higher levels of crop productivity than all other land tenure groups. It also varies due to who lease in and lease out land for cultivation. For instance, marginal and small farms leasing in land from resident land owners (T_4) were found to have comparatively higher gross productivity levels than marginal and small farms leasing in land from absentee land owners (T_3). But net returns per hectare for tenants were higher in the case of T_3. Further, net returns varied widely between various land tenure groups due to varying terms of lease. Under crop-sharing arrangements, tenants get relatively lower returns than those under the systems of fixed produce or fixed cash. Among pure tenants belonging to T_9 and T_{10} groups, the per hectare gross returns were more in the case of T_{10} because of close supervision and provision of credit and other inputs by the land owners. This confirms the Braverman-Stiglitz hypothesis that inter linkages of credit and tenancy contracts may serve landlords as a screening device to identify efficient tenants. But in that case, the extent of exploitation, also was more. In fact, as Bardhan (1980) rightly mentions in his theoretical formulation, various aspects of these relationships need more careful examination. However, productivity of tenant farms of all categories were found to be lower than both family farms (T_1) and capitalist farms (T_2) in all the villages studied.

What do the above findings indicate? Can we regulate tenancy arrangements in favour of more productive ones? Can we stipulate replacement of inefficient share cropping by a system of fixed rent? Besides, in view of the lower productivity of leased land, we may be promoted to suggest that abolition of tenancy would be more productive. But will it not lead to alienation of the poor from land, unless we record their right first and then confer ownership right to the tenants? Also, in the case of commercial tenancy, abolition of tenancy by conferring ownership right on rich tenants, may tilt the power balance of villages against the poor. These are some of the questions which need to be resolved. Moreover, in the wake of new economic policy of liberalisation, one may be tempted to argue that tenancy should be opened up for flexibility and growth. In this context, however, we need to keep in mind that all forms of tenancy should not be opened up either for growth or equity. We must be selective in our approach.

REFERENCES

Bardhan, P.K. and T.N. Srinivasan (1970). 'Crop Sharing Tenancy in Agriculture : A Theoretical and Empirical Analysis', *American Economic Review*, March. 1970, pp. 48-64.

Bardhan, Pranab K. (1980). 'Inter-Locking Factor Markets and Agrarian Development : A Review of Issues, *Oxford Economic Papers* 32(1), 1980.

Braverman, Avishay and Joseph E. Stiglitz, (1982). 'Share Cropping and the Interlinking Agrarian Markets', *American Economic Review*, Sept. 1982, pp. 695-715.

Cheung, S.N. (1969). 'Theory of Share Tenancy', Chicago University Press, 1969.

Cheung, S.N.S. (1968). 'Private Property Rights and Share Cropping', *Journal of Political Economy*, Vol. 2, November. 1968, pp. 1107-1122.

Haque, T. (1992). 'Economics of Agriculture in Backward Regions, with Special Reference to Viability and Sustainability of Small and Marginal Farms', NIRD. Hyderabad (mimeo).

Hsiao, J.C. (1975). 'The Theory of Share Tenancy Revisited', *Journal of Political Economy*, Vol. 83, Oct. 1985, pp. 1023-1032.

Marshall, Alfred 'Principles of Economics, 8th edition, London, Macmillan, 1956.

Newbery, D.M.G. (1977). 'Crop Sharing Tenancy in Agriculture : Comment'. *American Economic Review*, Vol. 64, Dec. 1974, pp. 1060-1080.

Newbery, D.M.G. (1977). 'Risk Sharing, Share Cropping and Uncertain Labour Markets', *Review of Economic Studies*, Vol. 44, Oct. 1977.

Reid, J.D. Jr. (1975). 'Share Cropping in History and Theory', *Agricultural History*. Vol. 49, 1975, pp. 426-440.

6

SUMMARY AND CONCLUSION

The study entitled 'Sustainability of small Holder Agriculture in India', was undertaken with a view to analysing the profitability of various agricultural enterprises and examining the viability and sustainability of small and marginal farms in the context of new economic policy of liberalisation and market reform. We ventured to examine these issues assuming that Indian Agriculture would continue to be dominated by small farms, unless the entry of multi-nationals and corporate sectors to agriculture squeeze them out in the long run for which there is a remote possibility. Besides, the roles of size of farm and land tenure system in influencing farm productivity, employment and occupational diversification were examined and appropriate policy measures for viability and sustainability of small holder agriculture in India have been suggested.

The study was based on collection and analysis of both secondary and primary data. A comprehensive primary level survey was conducted in the states of Andhra Pradesh, Bihar, Rajasthan and Haryana, selecting two representative districts from each state. Furthermore, a duly structured questionnaire schedule was canvassed in selected relatively advanced and relatively backward districts, covering a total of 900 farm households.

Cropping Patterns and Yields

The study revealed that there are large scale inter-state variations in cropping patterns, yields and factor productivity of crops in India. Since cash crops like sugarcane, cotton, coconut, potato etc. generate higher returns than the cereal crops like rice, wheat, sorghum, maize etc. it is often argued that changes in cropping pattern in favour of high value, albeit cash crops should form an essential component of any strategy of small farm development. However, not withstanding the merits of small farm diversification, the relatively very high yields of cereal crops like rice and wheat in Haryana and Punjab suggest that given the access to technology and other infrastructural and institutional facilities, even cereal crops have potentials which need to be explored and utilised properly. Moreover, small farmers prefer to grow cereals and pulses for the sake of food security and other conveniences. It is only in Kerala that plantation crops dominate the agricultural scene and sustain many small and marginal farms.

Profitability of Crop Enterprises

It is clearly borne out from the analysis of cost of cultivation data that by and large, there has been a declining trend in the net profitability of various crops over time. This is primarily due to rising prices of critical farm inputs like labour, fertilisers, pesticides etc. and not so much for increase in factor use or decline in marginal value productivity of these inputs. It could also be observed that both gross and net returns from rice in the traditional rice growing states of Andhra Pradesh, Assam, Bihar, Madhya Pradesh, Orissa and Uttar Pradesh and West Bengal, were comparatively lower than that of non-traditional rice growing states of Punjab and Haryana. Besides, there was lot of instability in the rate of returns. Generally speaking, both output-input ratio and factor productivity showed declining trends in the case of all the major cereal and pulse crops and also in the case of cash crops like Groundnut, Cotton and Jute. The factor-productivity (which indicates the net profitability of crop cultivation) showed rising trends only in the case of sugarcane and potato and a mixed trend of rise in some states and decline in some other states in the case of rape-seed and mustard. On the whole, however, there was a declining trend in factor productivity and consequently in the rate of profit of crop farming.

Variation in the Structure of Cost of Cultivation

During the past one and a half decade or so, there has been a significant change in the structure of cost of cultivation of crops in various regions. The percentage shares of new technical inputs in the total operational cost have increased in a number of states including Punjab and Haryana. Nevertheless, due to relatively greater rise in the wages of human labour, the percentage share of wages in the total operational cost was found to be highest in the case of all cereal crops excepting wheat.

In the case of rice, there were six broad patterns. First, in the state of Andhra Pradesh, the percentage share of wages in the total operational cost was as high as 41 per cent, followed by chemical fertilizer (21%), bullock labour and other expenses respectively. Second, in the relatively developed states of Haryana and Punjab, the next to cost of human labour was that of irrigation, followed by chemical fertilizers, machine labour and other items respectively. Also a substantial amount of expenditure was incurred on plant protection measures in these states. Third, in the relatively under developed regions of Bihar, Madhya Pradesh, Karnataka, Orissa and West Bengal, the highest percentage share of cost of human labour (wage) was followed by expenditure on the use of bullock labour, chemical fertilizers and other items in that order. Fourth, in the state of Tamil Nadu, the cost on human labour was followed by that of chemical fertilizers, bullock labour, irrigation and other items respectively. Fifth, in the state of Uttar Pradesh, the percentage share of human labour in the total factor cost was nearly 30 per cent, followed by bullock labour (12%), chemical fertilizers (7.4%) and other items of cost respectively. Finally, the state of Assam had a unique traditional cost pattern in the sense that cost of human labour was followed by that of bullock labour and seeds instead of new technical inputs.

In the case of wheat, the impact of new technology on the cost structure was more pronounced. In the states of Punjab and Haryana, the expenditure on chemical fertilizers accounted for the highest percentage share in the total operational cost followed by human labour, machine labour, seed, irrigation charges and other items of cost respectively. In the case of Punjab, the percentage share of human labour in the total factor cost declined from 31 per cent in 1973-74 to 27 per cent in 1988-89, while that of machine labour increased from 11.3 per cent in 1973-74 to 22.9 per cent in 1988-89. However, in the states of Madhya Pradesh, Rajasthan and Uttar Pradesh a traditional cost pattern with human labour being followed by bullock labour could be found. This was also true in the case of other cereal crops like Jowar, Maize and Bajra, presumably because of low level of technological change in the case of these coarse cereal crops.

Profitability of Crop Enterprises as Revealed through Farm Level Survey

The results of our farm level survey in the year 1990-91 were quite revealing. Per hectare as well as per farm productivity varied widely from region to region. The annual net returns per hectare of net cultivated area varied between Rs. 1091 in Chenampally (Anantapur) to Rs. 13,992 in Ponangi (West Godavari). In Anantapur, small and medium farms had relatively higher returns per unit of net cultivated area than both marginal and small farms mainly because of their more intensive use of land and labour. But in the relatively developed district of West Godavari where farming was highly capital intensive, large farms in Ponangi village and medium farms in Vatlure village had an edge over all other categories of farms. Besides, due to low size of holdings, annual net returns per marginal and small farms were comparatively very low in both developed and under-developed villages of Andhra Pradesh.

In Bihar, small and medium farms had relatively higher returns than marginal and large farms in all the four villages of Bhagalpur and Patna districts studied. The net crop income per marginal farm ranged from Rs. 1053 in Mathurapur (Bhagalpur) to Rs. 1994 in Simra (Patna) and that of small farms varied between Rs. 4642 in Mathurapur and Rs. 10237 in Simra. It is thus evident that net crop income of marginal and small farms were below the subsistence level or what is known as poverty line of Rs. 11000 per annum. In Haryana, the net returns per hectare of net cultivated area ranged from Rs. 3026 on large farms in Chandabass (Bhiwani) to Rs. 9985 on medium farms in Gagsina. The net crop income of marginal farms ranged between Rs. 1336 in Chandabass II to Rs. 5133 in Gagsina and that of small farms varied from Rs. 4006 in Chandabass II to Rs. 15541 in Gagsina (Karnal). Thus, even in Haryana, it was only in the relatively developed village of Karnal district that small farms were found to be above poverty line. In Rajasthan also, the average crop income of marginal and small farms were much below the subsistence level in all the villages studied, excepting the irrigated village of Netewala in Sriganganagar district where small farms with an average size of 1.46 hectare of land realised a net crop income of Rs. 14836 per annum. On the whole, our farm level survey clearly bears out that without adequate access to land, technology and institutional facilities of credit, market etc., crop based small and marginal farmers would be forced to live below the subsistence level.

Economics of Horticulture

In view of low net returns from the cultivation of traditional crops, it is often suggested that diversification in favour of horticulture, sericulture, acquaculture etc. would hold the key to viability and sustainability of farms. It is particularly emphasised that there is tremendous potential for increasing exports of fruits and vegetables by India because of her comparative advantage in labour intensive horticultural products. In recent years, the export value of fruits and vegetables in India increased from Rs. 180 crores in 1990-91 to Rs. 456 crores in 1994-95.

However, in our study villages, horticulture was not found to have been adopted by farmers on commercial scale, although some large farmers of Anantapur villages were cultivating mango crop quite profitably. It was observed that one hectare of malguba variety of mango cultivation yielded a net income of Rs. 25339 per annum which is difficult to realise from other traditional crops. But there are some practical difficulties in the adoption of mango cultivation by small and marginal farms. First, there is gestation period of 2 to 3 years before mango plants start giving fruits and during this gestation period, marginal and small farmers need some alternative sources of income for subsistence. Besides, without adequate institutional credit facilities, small and marginal farmers would not be able to invest about Rs. 12000 in order to realise a net income of Rs. 25339 per hectare. Moreover, unless the villages are connected with road and market, middlemen would tend to exploit the farmers and reduce their profit margins. Furthermore, any substantial increase in the area under fruits in the near future is unlikely because of lack of irrigation water/soil moisture in dry regions and absence of high quality planting material, processing units, infrastructural support for packaging, storage, cooling and transport. It is often suggested that a system of contract farming in which small farmers produce fruits and vegetables for co-operative or corporate bodies under contractual arrangement may help remove some of these constraints to small farm diversification.

Economics of Sericulture

Similarly, mulberry cultivation is said to hold lot of promises for generation of income and employment for the small and marginal farmers. In one of our study villages of Anantapur district, per hectare net return from mulberry cultivation (as of 1990-91) ranged from Rs. 6000 to Rs. 15,000. There was large scale inter-farm variation in returns, due mainly to crop loss for diseases on some farms and here lies the risk of mulberry cultivation by small and marginal farmers. The co-efficient of inter-farm variations in gross returns was as high as

33.1 per cent. In fact, some farmers realised negative returns from mulberry cultivation which acts as a disincentive to the farmers to adopt sericulture on a large scale. Unless there is an effective policy to provide insurance against risk of crop failure and an institutional mechanism for the supply of cross bred disease free layings, small and marginal farmers may not easily be motivated to switch over to mulberry cultivation. Moreover, improved institutional credit and marketing facilities, water management techniques and development of watersheds would be necessary. Nevertheless, in view of its high employment and income generating capacity, all out efforts should be made to ensure the above mentioned facilities for wider adoption of mulberry cultivation, wherever feasible.

Economics of Animal Husbandry

Animal Husbandry is considered to be one of the important means of livelihood for many farm households in India. In recent years, production of both milk and eggs have significantly increased. However, the relatively backward regions of the country have yet to witness sufficient progress in milk production. In our study villages, the average number of milch animals per farm varied from 1 to 5 in various size groups of farms in different selected regions of the country. While in the relatively developed district of West Godavari, marginal and small farms were found to maintain relatively greater number of milch animals than the large farmers; in the dry, albeit under developed district of Anantapur, the average number of milch animals maintained by marginal and small farms were comparatively less. But in both West Godavari and Anantapur Districts, marginal and small farmers were found to maintain milch animals mostly on lease/milk sharing basis. In such cases, 50 per cent of the milk were to be paid to the cattle owner, while the entire cost of maintanance was borne by the tenants. In the remaining study areas, no such incidence of leasing of milch animals was reported.

The results of regression analysis further showed that there was a positive relationship between farm size and gross as well as net returns per milch animals in most places which could be explained by large farmer's access to improved feed for animals, including green fodder, dry fodder and concentrates. The small and marginal farms do have neither enough land for the purpose of growing fodder nor adequate purchasing power to buy fodder and concentrates from others. Generally speaking, they were found to depend on common grazing lands, pastures and fodder supply by their village patrons and remain satisfied with low milk yields and returns. Moreover, in places where public animal health care facilities were not available, small and marginal farmers found it difficult to maintain cross-bred, albeit high yielding milch animals due to their more proneness to diseases. Besides, in under-developed villages, due to lack of market, marginal farmers who were dependent on large farmers for employment and credit, tended to sell their milk produce to the patrons at relatively lower prices. In addition, small and marginal farmers who maintained milch cattle mainly on lease/milk sharing basis, earned lower net returns per animals, because they had to part away with 50 per cent of the milk produce to the cattle owners. The erosion of common grazing lands in recent years has further reduced the small and marginal farmer's ability and incentive to maintain milch cattle on profitable basis.

Economics of Fish Cultivation

Fish cultivation is considered to be one of the important ways of occupational diversification and sources of additional income and employment for the farmers. It is particularly emphasised at the research station level that integrated farming system involving paddy cum fish culture, fish-duck integration etc. can become an effective source of additional income generation. Unfortunately, this potential remained unutilised thus far. In most of our study villages, fish cultivation was more or less non-existent. But in Ponangi village of West Godavari District, quite a few large and medium farmers has adopted fish culture which yielded an average income of Rs. 39,040 per hectare. But small and marginal farmer's inadequate access to land and institutional marketing and credit facilities stood in the way of their adopting fish cultivation on commercial scale. In addition, a change from crop cultivation to fish culture involved certain risks and uncertainties which most of the marginal and small farmers tended to avert due to their low capacity to bear risk of production/income loss arising out of fish diseases, theft etc.

Farm Size and Productivity

In view of the conflicting results of various studies, it was considered necessary to re-examine various issues relating to size-productivity relationship in greater details. It was observed that there was no uniform pattern of relationship between farm size and crop yields in various regions of the country. Still certain trends were evident. First, marginal as well as large farms produced less per unit of total cropped area than small and medium farms, although the reasons for low crop yields of marginal farms could be different than those of large farms. While marginal farms suffered due to capital and other input constraints, large farms lacked necessary managerial ability and incentives to produce more. Secondly, in most of the technologically developed areas, there was an emerging trend of medium size farms (5 to 10 acres) to produce more per unit area than all other categories of farms. This does invariably suggest that traditional inverse relationship between size of farm and crop productivity tended to weaken in technologically developed areas. But the very fact that large farms produced less per unit of cropped area than medium farms in technologically developed villages and as compared to both small and medium farms in all other villages demonstrate the necessity of an appropriate size of farm, not unmanageably too large. In six out of fifteen villages studied, there was an inverse relationship between farm size and gross returns per hectare, while positive relationship between farm size and gross returns per hectare could be observed in the remaining villages. Since some of the technologically developed villages also showed strong inverse relationship between farm size and productivity, it proves beyond doubt that under favourable environment, small and marginal farms could adopt new technology and positive interaction between technology and high intensity of land and labour uses would result in higher yields on such farms.

Need for Land Reform

An economic case for land reform has generally been made on the basis of size productivity relationship. Since our survey results showed that small and medium farms raise higher yields per unit of area than both large and marginal farms, it may be desirable to take away land from very large holdings for their distribution among marginal farms so that marginal farms could be converted into small and viable ones. But given the pressure of population on land, it is likely that most of our farms would tend to become marginalised in the near future and without sufficient technological breakthrough in agriculture and development of non-farm employment, land reforms alone may fail to yield the desired result. It is true that changes in the rural power structure through land reform would trigger off certain positive forces of growth for rapid cum balanced rural development. But given the present unfavourable socio-political trends in the country, we could only say that this is a long drawn process.

However, both marginal and large farms were found to have greater potentials for migration outside agriculture than small and medium farms who were more attached to land. While marginal farmers tend to migrate because of non-viability of their farms and economic compulsion to earn additional income by hiring out labour in neighbouring towns/factories, the large farms, if forced to migrate, would depend for income creation on their entrepreneurial abilities rather than landed assets. This kind of logic further strengthens the economic case for land redistribution. Besides, sharecropping turned out to be most inefficient form of land tenure whose replacement by a system of fixed rent or ownership cultivation might help increase farm productivity.

Viability and Sustainability of Marginal and Small Farms

Notwithstanding the above mentioned arguments for land reform, the question of viability and sustainability of marginal and small farms assumed special significance, particularly in the context of agriculturally backward regions. In our study areas, marginal farms could no where (including the developed districts of West Godavari, Karnal, Patna and Sriganganagar) generate sufficient income from crop farming alone to stay above poverty line. Of course, if we judge viability on the basis of efficiency measured in terms of profit function, then all size groups of farms were found to be viable in our study villages. But if viability is defined on the basis of minimum level of income for meeting the consumption plus other precautionary and investment needs of farms, then marginal farmer's crop income were found to be far below that level everywhere. Even small farms could generate crop income above the subsistence level only in the relatively developed villages of Ponangi, Vatlure (West Godavari), Gagsina (Karnal) and Netewala (Sriganganagar). In all other villages, net crop income of small farms were much below the nationally accepted subsistence level of Rs. 11000. It is clearly borne out from our analysis that in the relatively backward villages of Anantapur, even the medium farms do not generate enough crop income to stay above subsistence level and large farms earn hardly subsistence level of income through crop farming. It is thus clear that by and large, marginal and small farms cannot earn a viable level of income from crop farming alone. This is particularly true for crop farming under dry and rainfed conditions. In our study villages, marginal and small farmers depended mainly on other sources of income for their subsistence. These included leasing in land, animal husbandry, poultry, wage paid employment etc. But even when we include income from all other sources, marginal farmers were found to be below the subsistence level in most of the villages. Similar was the case with small farms in unirrigated areas. Thus, when marginal and small farms were not viable on the basis of subsistence level of income criterion during a normal agricultural year, their sustainability under adverse circumstances like drought, flood etc. becomes all the more uncertain. Also their capacity to sustain gets reduced due to continuous soil degradation, erosion of common property resources and ever increasing pressure of population on land. Moreover, a farm household needs income not only for meeting its consumption requirements, but also for facing emergency situation like illness, accident, drought, flood etc. In addition, there should be some savings for investment for future growth. If we assume that at current prices, Rs. 28,000 per farm is a reasonable level of income for some such sustenance (as this is defined as the income tax exemption limit in urban areas), then even large farms in many dry zones do not generate sufficient income from crop farming alone for the purpose. The minimum size of holding required to generate a threshold level of crop income under existing situation (based on subsistence level of income) ranged from 0.8 hectare in Ponangi village of West Godavari to 9.1 hectare in Chenampally village of Anantapur district. But on the basis of Rs. 28,000 per farm criterion, the minimum required size of land holding ranged from 2.1 hectare in Ponangi village (West Godavari) to 23.1 hectare in Chenampally village of Anantapur. In fact, the main argument of those who oppose low ceilings on land holdings in dry regions is that even medium and large farms do not generate sufficient income for sustenance.

However, viability and sustainability of farms should not be considered as static concepts. It varies due to changes in technology, cropping patterns, farming system, soil health, input-output prices, population growth, infrastructural facilities and possibilities of integration between farm and non-farm activities. Moreover, what is found to be viable to-day, may or may not remain so in future depending on factor-productivity, factor-relations and extent of population pressure on land. In fact, given the present trend of population pressure on land to continue, most of the marginal households may join the rank of landless labourers soon which in the absence of sufficient growth of farm and off-farm employment opportunities, may pose a threat to socio-economic stability in the villages. To the extent, socio-economic environment gets disturbed due to growing landlessness and unemployment, the agrarian system as such looks non-viable and non-sustainable.

Role of Technological Change

It is evident from the analysis of our farm level data that small farms in irrigated, albeit technologically developed areas were quite viable in the sense of both efficiency of production and income levels. But dry regions depicted a very depressing picture where due to non-availability of appropriate dry farming technology and low use of new technical inputs, even large farms failed to generate necessary income for sustainable development. The analysis of ICAR's data of national demonstrations further show that there are large technological/yield gaps in some of the backward regions which if fully exploited, may help enable the marginal and small farms to generate sufficient income above the subsistence level. However, it also depends on crop-rotation, cropping patterns and also yield response to technology adoption. In fact, yield response to technology adoption in unirrigated/rainfed areas were much lower than those of irrigated areas. Irrigated land capable of producing two or more crops in a year, using recommended level of production technology may sustain small and marginal farms on viable basis. But in unirrigated/dry regions where farmers are able to grow only one crop with very low yield response to available technology, there is remote possibility of small and marginal farms becoming viable, unless there is a sufficient breakthrough in dry farming technology or a shift from traditional cereal crops to high value crops including fruits, vegetables, sericulture etc. Furthermore, such crop or non-crop based diversification would require appropriate technological and infrastructural/organisational support for credit, marketing, storage and transportation. Besides, the provision of insurance would be important particularly because of the frequent occurrence of crop loss due to drought/flood, plant diseases etc., the risks of which must be covered for the sustenance of marginal and small farms. Moreover, in view of rising prices of chemical fertilizers as well as environmental hazards associated with chemical nutrients, it would be essential to develop and promote the use of low cost bio-fertilizers/natural farming among small and marginal farmers for sustainable development.

Impact of New Economic Policy of Liberalisation on Small Farm Development

Under the impact of new economic policy of liberalisation, currently there is an emphasis on export led agricultural growth through integration of our village economy with the rest of the world. While such trade linkages may have some advantages, this cannot be relied upon as an efficient strategy of development. Since it looks more to markets abroad, while ignoring the need to develop domestic market for growth, it is likely that necessary thrusts for employment oriented agricultural growth through land development and appropriate technological and infrastructural changes would be missing. Moreover, any major shift in the cropping pattern in favour of fruits, vegetables and other high value crops may not attract the small and marginal farmers, unless foodgrains are made available to them at subsidized prices and various land, capital and market constraints are removed. Contract farming may appear to be a solution. But it has its own advantages and disadvantages.

To sum up, small and marginal farmers may respond to the needs of commercial agriculture on viable basis only if their accessibility to appropriate technology, credit, insurance, market, price/subsidy and extension get sufficiently enhanced. For this, investment priorities may have to be reformulated. First, there should be renewed, albeit greater emphasis on yield increasing and cost reducing technological change for growth of agricultural income which in turn would stimulate demand for the growth of rural diversification and industrialisation. The relatively backward, albeit dry regions of the country are highly starved off appropriate technology with respect to both traditional cereal crops and non-traditional high value crops. This problem needs to be solved with critical minimum levels of investment and efforts in agricultural research and extension in appropriate direction. Second, development of infrastructure would be essential for the rapid growth of both agriculture and industry. Moreover, small and marginal farms in backward regions would need a little more special care and attention in terms of delivery services and integration of farm and non-farm activities. Let the proposed agri-business consortium help the small and marginal farms in solving some of these problems for their sustainable development.

Policy Conclusions

The Policy Conclusions drawn on the basis of this study may be summarised as follows:

1. Changes in cropping pattern in favour of high value crops, coupled with appropriate integration of farm and non-farm activities should form an essential component of any strategy of small farm development. This needs to be achieved through various technological, infra-structural, institutional and organisational changes. There is no pan India solution. However, depending on the need and suitability of each agro-climatic zone, small and marginal farmers may be encouraged to participate in co-operative or even contract farming arrangement with agro-processing units, without endangering their security of land ownership and land tenure.
2. Both output-input ratio and factor productivity showed declining trends in the case of all major cereal and pulse crops and also in the case of cash crops like groundnut, cotton and jute. Factor productivity showed rising trends only in the case of sugarcane and potato and a mixed trend of rise in some states and decline in some other states in the case of rapeseed and mustard. This requires greater research efforts for generation of yield argumenting, albeit cost reducing technologies for various crops including cereals, pulses, and horticultural crops.
3. Horticulture as well as sericulture were found to yield large returns. The future pattern of demand driven agricultural growth would depend largely on horticulture revolution. But there are some practical difficulties in the adoption of horticulture by small and marginal farms which include (i) low size of land holding; (ii) long gestation period; (iii) lack of high yielding technology; (iv) lack of institutional credit, marketing, storage and transport facilities; In the case of sericulture also, non-availability of cross bred, albeit disease free layings act as a major constraint. Removal of these constrains should receive attention on priority basis.
4. The impact of new economic policy of liberalisation on agricultural development may be non-significant, unless small and marginal farmers respond to the needs of commercial agriculture through their increased accessibility to appropriate technology, credit, insurance, market and linkages with non-farm employment opportunities.
5. There are huge untapped technological potentials in many regions which if fully exploited, may help enable the marginal

and small farmers to generate sufficient income above the subsistence level. However, in view of rising prices of chemical fertilizers as well as environmental hazards associated with chemical nutrients, it would be essential to promote the use of low cost bio-fertilizers and vermiculture among small and marginal farms for sustainable development.

6. The traditional agricultural development policies in India have not been favourable to diversified agricultural growth, particularly in the context of small farms and backward regions. First, agricultural price policy does not provide protection in terms of Procurement and minimum support prices to fruits and vegetable crops. Since the production and prices of these commodities fluctuate widely, small farmers would find it difficult to grow these crops on commercial scale, unless price protection is available. Secondly, small farmers in backward regions do not have access to cold storage, market and transportation facilities and consequently, the promotion of agro-processing activities remain a dream. Primarily, it is the responsibility of the government to develop infrastructural facilities, although willing private entrepreneurs or community efforts of farmers would be welcome. Thirdly, tenancy laws of most of the states may have to be modified to allow the large farmers to take up non-farm activities and to enable the marginal farmers to have greater access to land through lease market. This would enhance their economic viability and also promote diversified agricultural growth. In view of high concentration of land holdings in the hands of a few cultivators in each region on the one hand and growing marginalisation of small farms on the other, there appears to be a strong case for further redistribution of land. If it is not possible to do so, at least the government should give priority to development of wastelands and their distribution among the landless and semi-landless people. Similarly, consolidation of holdings of even smaller holdings would enable them to enjoy better economies of scale. Fourthly, in view of high variability of production, prices and income of so-called high value agricultural commodities, the possibility of crop insurance coverage may be considered, particularly for those commodities which are perishable, but of great social and economic value.

BIBLIOGRAPHY

Acharya, M.S. (1993), Irrigation Methods for Water Deficient Areas; Farmer and Parliament, 28(11), 1993 (November); 17-18.

Acharya, S.S. and Sunita Verma (1990), Viability and Sustainability of Agriculture in Rajasthan *in* Proceedings of the National Seminar held at NIRD, Hyderabad, October 8-10, 1990.

Alagh, Y.K. (1990), Agro-climatic Planning and Regional Development; *Indian Journal of Agricultural Economics*, July-Sept., 1990; pp. 244-268.

Ahmed, Nisar (1990), Lending Policy of Co-operative Credit Institutions and Development of Agriculture; *Co-operator*, 27(16), 1990 (February 15); 435-436 & 446.

Allexandra Bonanno (1989), Sociology of Agriculture; Technology, Labour, Development & Social Classes in an International Perspective; Concept Publishing Company, New Delhi, 1989.

Ali Mohammad (1985), Dynamics of Agricultural Development in India; Concept Publishing Company, New Delhi.

Anne W. Van den Ban and Arnold L.G.M. Bahwens (1986), Small Farmer Development; Experiences in the Nether Lands, *Quarterly Journal of International Agriculture*, Vol. 27, July-December, 1986, 216-227.

Bal, H.S.; Jain, K.K. and Bant Singh (1991), An Economic Analysis of Farm business in Punjab; *Agricultural Situation in India*, July, 1991, pp. 191-201.

Bandyopadhyay, Rekha (1994), Returning Agrarian Reform to Centre-Stage; *Economic and Political Weekly*, 29(12), 1994 (March 1-9); pp. 681-682.

Bandyopadhyay, S.C. (1993), Impact of New Economic Policy on Agricultural Investment; *State Bank of India, Monthly Review*; 32(4), 1993 (April), 176-187.

Banerjee, Amalesh (1993), The Impact of New Economic Policy on Agricultural Labour; *Indian Journal of Agricultural Economics*; 48(3), 1993 (July-Sept.); 471-476.

Banerjee, Biswanath and Roy, Soumendra Dutta (1987). Economics of Small and Large Farms in Haringhata Block of Nadia district, West Bengal—A Case Study; *Economic Affairs,* 32(1), 1987 (March); 31-37.

Bardhan, Pranab K. (1980), Inter Locking Factor Markets and Agrarian Development—A review of issues, Oxford Economic Papers, 32(1); 1980.

Bardhan, P.K. and T.N. Srinivasan (1971), Crop-Sharing Tenancy in Agriculture—A Theoretical and Empirical Analysis; *American Economic Review*; Vol. 69, March, 1971; pp. 48-64.

Bhaduri, Amit (1973), A Study in Agricultural Backwardness under Semi-Feudalism; *The Economic Journal*, Vol. 83, No. 329, March 1973.

Bhagwati, J.N. and Chakravarthy, S. (1969), Contributions to Indian Economic Analysis—A Survey; *American Economic Review Supplement*, September, 1969.

Bhalla, G.S., N.S. Randhawa and D.S. Tyagi (1989), Implementation and Administration of Agricultural Price Policy in Asia; FAO—Rome; 1989.

Bhalla, G.S. in Berry & Cline (1979), Agrarian Structure and Productivity in Developing Countries; John Hopkins University Press, Baltimore, U.S.A., 1979.

Bhalla, Sheila (1989), On the Proliferation of Small Farms and Labour Absorption in Indian Agriculture; *Man and Development*, 11(3), 1989 (September); pp. 94-112.

Bhardwaj, Krishna (1974), Notes on Farm Size and Productivity; *Economic and Political Weekly*; Review of Agriculture, March 30, 1974.

Bhattacharya, N. and G.R. Saini (1972), Farm Size and Productivity—A Fresh Look; *Economic and Political Weekly*; June 24, 1972.

Bigman Maloa, M.J. and Nkosi, S.A. (1993), Agricultural Development Through Contract Agents; Appropriate for Small Holders?; *Development Southern Africa*, 10(4), 1993; pp. 515-534.

Binswanger, Hans and J.B. Quizon (1986), What can Agriculture do to the Poorest Rural Groups?; ARU Report No. 57, The World Bank, 1986.

Binswanger, Hans and M.R. Rosenweig (1979), Contractual Arrangements, Employment and Wages in Rural Labour Markets; A critical review, New York, ADC.

Binswanger, H.P. and M.R. Rosenzweig (1979), Behaviourial and Material Determinants of Production Relations in Agriculture; *Journal of Development Studies*, Vol. XXII, No. 3, pp. 503-539.

Blake, Robert O. (1992), Sustainable Agriculture on Small Farms—Time to Move Forward; *Indian Farming*; 42(2), 1992.

Braverman, Avishay and Joseph E. Stiglitz (1982), Share Cropping and the Interlinking Agrarian Markets; *American Economic Review,* September, 1982; pp. 695-715.

Bunyasi, Sakwa (1985), Agricultural Policy and its Relationship to Food Policy in Sub-Saharan Africa; Report on Two Seminars....11-22, March, 1985. Washington; Economic Development Institute of the World Bank, 1985.

Cakmak, Erol H. (1992), Medium-term Growth Prospects for Turkish Agriculture; A Sector Model Approach; *Developing Economies*, 30(2), 1992 (June); pp. 132-153.

Chand, Ramesh (1991), Agricultural Development, Price Policy and Marketed Surplus in India : Study of Green Revolution Region; Concept Publishing Company, New Delhi, 1991.

Chand, Ramesh and S.C. Tewari (1991), Growth and Instability in Indian Exports and Imports of Agricultural Commodities; *Indian Journal of Agricultural Economics*, April-June, 1991; pp. 159-165.

Chattopadhaya, M. and Ashok Rudra (1976), Size—Productivity Revisited, *Economic and Political Weekly*, Review of Agriculture, September, 1976; pp. A104-106.

Chatterjee, T. (1988), Net Returns Optimisation Through Resource Efficiency Analysis of Crop-Mixes; An Experiment for Rayalseema Small Farms; *Indian Journal of Agricultural Economics*, 43(1), 1988 (January-March); pp. 63-69.

Chaudhri, D.P. (1993), Productivity Trends in Asian Agriculture, 1950-1990; *Indian Journal of Labour Economics*, 36(2), 1993 (April-June); pp. 169-189.

Cheung, S.N.S. (1968), Private Property Rights and Share Cropping; *Journal of Political Economy*, Vol. 76, 1968; pp. 1107-1122.

Cheung, S.N.S. (1969), Theory of Share Tenancy, Chicago, University Press, 1969.

CIRDAP, FAO—Commonwealth Secretariat (1992), Report of the Regional Workshop on 'Organisation and Management of Agricultural Services for Small Farmers in 'Asia', Dhaka, Bangladesh; June 28-July 2, 1992.

Dantwala, M.L. (1994), Agriculture Policy; Prices and Public Distribution System—A Review; *Manthan*, 15(1), 1994 (Jan.-March); pp. 73-91.

Dantwala, M.L. (1991), Agrarian Structure and Agrarian Relations in India *in* India's Agricultural Development Policy, Oxford-IBH Publishing Co., New Delhi, 1991.

De, Aitabha (1993), Institutional Impact on Farm Production, Investment and Income—An Indian Case Study; *Indian Cooperative Review*, 31(1), 1993 (July); pp. 79-89.

Dasgupta, Biplab (1980), The New Agrarian Technology and India; *Macmillan*, Delhi, 1980.

Deoghare, P.R. and Sharma, B.M. (1992), Prospects of Increasing Farm Income on Marginal and Small Farms; *Agricultural Situation in India*, 46(12), 1992 (March), pp. 925-929.

Deolalikar, Anil B. (1981), The Inverse Relationship Between Productivity and Farm Size—A Test Using Regional Data from India; *American Journal of Agril. Economics*, Vol. 63, No. 2, May, 1981, pp. 275-79.

Desai, Bhupat M. (1992), Institutional Finance for Agricultural Development; an analytical survey of critical issues; Oxford-IBH Publishing Company, New Delhi, 1992.

Don Kanel (1967), Size of Farm and Economic Development; *Indian Journal of Agricultural Economics*, Vol. XXII, No. 2, Apr.-Jun., 1967, pp. 26-43.

Dreze, J.; Famine Prevention in India, Discussion Paper Series No. 3, London School of Economics.

Dubashi, P.R. (1986), Policy and performance : Agricultural and Rural Development in Post-Independence India, New Delhi; Sage Publications, 1986; pp. 320.

Due, J.M. and Magayane, F. (1990), Changes Needed in Agricultural Policy for Female-Headed Farm Families in Tropical Africa; *Agricultural Economics*, 4(3/4), 1990 (December); pp. 239-253.

Ehui, Simon K. and Spencer, Dunstan, S.C. (1993), Measuring the Sustainability and Economic Viability of Tropical Farming Systems : A model from Sub-Saharan Africa; *Agricultural Economics*, 9(4), 1993 (Dec.); pp. 279-296.

Ervin, David E. (1993), Trade Agreements, Agriculture and the Environment in Developing Countries; *American Journal of Agricultural Economics*, 75(3), 1993 (Aug.); pp. 799-800.

Floris P. Blankenberg (1991), Appropriate Technology for Rural Development in India; Concept Publishing Company, New Delhi, 1991.

Food and Agriculture Organisation (1988), WCARRD, Ten Years of Follow-up; 1979-89; Rome; 1988.

Food and Agriculture Organisation (1979), Regional Office, Bangkok; Rural Markets : A Critical Link for Small Farmer Development—A Country Report and Case Studies on Rural Markets in 10 Asian Countries; 1979.

Food and Agriculture Organisation (1989), World Conference on Agrarian Reform and Rural Development : Report of the Sixth Government Consultation for Asia and the Pacific Region, (RAFA 1989/7), pp. 21-22.

Frankel, Francine (1978), India's Political Economy, Oxford University Press, Delhi, 1978.

Gandhi, B.M. and Ramachandran, S. (1992), Environmentally Sound Biotechnology; *Indian Farming*, 42(3), 1992 (Jun.); 15-26.

George, M.V., N.J. Kurian and Mohan C. Chandra (1983), Factor Shares in Indian Agriculture, Temporal and Spatial Variations and their Implication; *Indian Journal of Agricultural Economics*, Vol. XXXVIII, No. 3, July-Sept., 1983.

Ghose, Arabinda (1993), Draft Agriculture Policy Resolution and Agricultural Development; *Yojana*, 37 (14&15), 1993 (August 15); pp. 72-76.

Ghosh, Gautam and Kaur, Dilbag (1992), Counteracting the Constraints in the Transfer of Agrotechnologies; *Invention Intelligence*, 38(8), 1992 (Aug.); pp. 263-65.

Ghosh, S.P. (1993), Research-Extension Linkage in Agro-Climatic Zones; *Intensive Agriculture*, 31(1-2), 1993 (March-April); pp. 3-7.

Gupta, Jancy (1993), Scientific Prawn culture in Homesteads—A gainful self-employment opportunity for coastal women; *Indian Farming*, 43(5), 1993 (Aug.); pp. 25-27.

Ghonemy, M. Raid El. (1990), The Political Economy of Rural Poverty—The Case for Land Reform, Routledge, London and New York, 1990.

Gil Sucha Singh (1993), Dunkel Draft Text and Some Issues Related to India's Policy for Agricultural Development; *Indian Economic Panorama*, 3(2), 1993 (July); pp. 9-12.

Government of India (1990), Ministry of Agriculture, Report of the High Powered Committee on Agricultural Policies and Programmes, July, 1990; pp. 18-27.

Grover, D.K. and Others (1992), An Economic Analysis of Milk Production in Bhatinda District of Punjab; *Indian Journal of Dairy Science*, 45(8), 1992 (Aug.); pp. 409-415.

Gulati, Ashok and Others. Export Competitiveness of selected Agricultural Commodities, NCAER, 1994.

Haque, T. and A.S. Sirohi (1986), Agrarian Reforms and Institutional Changes in Agriculture in India: Concept Publishing Company, New Delhi, 1986.

Haque, T. and Earl. O. Heady (1982), Problem of Resource Allocation in Indian Agriculture, Centre for Agriculture and Rural Development; Iowa State University, U.S. (mimeo), 1982.

Haque, T. (1987), Temporal and Regional Variations in Agrarian Structure in India; *Indian Journal of Agricultural Economics*, July-Sept., 1987.

Haque, T. and C.C. Maji (1978), Structure and Flows of Agriculture Credit in India; *Indian Journal of Agricultural Economics*, Oct.-Dec., 1978, Vol. XXXIII, No. 4, pp. 72-78.

Haque, T. (1988), Role of Technological Change in Improving the Economic Condition of Marginal Farmers and Landless Labourers, *in* Proceedings of the National Seminar, Organised by IASSI, New Delhi, November 1-3, 1988.

Haque, T. and Sunita Verma (1988), Regional and Class Disparities in the Flow of Institutional Credit in India; *Indian Journal of Agricultural Economics*, July-Sept., 1988.

Haque, T. and G. Partha Sarathy (1992), Land Reform and Rural Development; *Economic and Political Weekly*, Vol. XXVII, No. 8, February, 1992.

Haque, T., V.K. Sharma and M.S. Bhatia (1983), Temporal and Spatial Variations in Factor Shares in Indian Agriculture; *Indian Journal of Agricultural Economics*, Vol. XXXVIII, July-Sept., 1983, pp. 391-399.

Harriss, Barbara; State and Market : State Intervention in Agricultural Exchange in a Dry Region of Tamil Nadu, South India; Concept Publishing Company, New Delhi.

Hazel, Peter B.R. and Steven Haggblade; Rural-Urban Growth Linkages in India; *Indian Journal of Agril. Economics*, Oct.-Dec., 1991, pp. 515-539.

Hazra, C.R. (1992), Watershed a Way to Sustainable Agriculture; *Intensive Agriculture*, 30(9&10), 1992 (Nov.-Dec.); pp. 11-17 and 39.

Hopper, W. David (1965), Allocation Efficiency in Traditional Indian Agriculture; *Journal of Farm Economics*, Vol. 47, No. 3, August, 1965; pp. 611-624.

Hopper, W. David (1975), Allocation Efficiency in Traditional Indian Agriculture; *Journal of Farm Economics*, 1975.

Heady, E.O. (1947), Economics of Farm Leasing System; *Journal of Farm Economics*, Vol. XXIX, Aug., 1947, pp. 659-78.

Hiromitsu Kaneda (1980), Structural Change and Policy Response in Japanese Agriculture After the Land Reforms; *Economic Development and Cultural Change*, Vol. 28(3), 1980.

Hsiao, J.C. (1975), The Theory of Share Tenancy Revisited; *Journal of Political Economy*, Vol. 83, Oct., 1975, pp. 1023-1032.

Huggar, L.B. and Others (1992), Constraints in Diffusion of New Farm Technology—An Economic Evaluation; *National Bank Review*, 8(7), 1992 (Sept.), pp. 25-30.

ICAR (1987). Annual Report on National Demonstration, New Delhi, 1986-87.

International Association of Agricultural Economists (1991), Proceedings of the XXI Conference of the IAAE, Tokyo, 1991.

International Rice Research Institute (1981). Los Bonos Consequences of Small Farm Mechanization on Rural Employment, Incomes and Production, in selected countries of Asia, Report of a Joint ADC-IRRI Workshop 14-18 September, 1981—Los Banos, 1981, iii, 50p.

I.P., P.C. and C.W. Stahl (1978), System of Land Tenure, Allocative Efficiency and Economic Development; *American Journal of Agricultural Economics*, Vol. 60, No. 1, 1978, pp. 19-28.

Jain, K.K. (1993), An Appraisal of Costs and Prices on Different Farms in Punjab; *Agricultural Situation in India*, 27(10), 1993 (Jan.); pp. 763-73.

Jain, H.K. (1994), The GATT proposals on Agriculture-Scientific Implication; *Indian Farming*, 44(1), 1994 (April); pp. 3-6.

Jasekharan, P.R. (1992), Agricultural Holdings and Off-farm Income of the Rubber Growers—A Farm Family Approach; *National Bank News Review*, 8(4), 1992 (June); pp. 24-26.

Jayaraj, S. (1993), An Analysis of Export Performance of Foreign Controlled Companies in India During 1979-80 to 1987-88; *Indian Economic Panorama*, 3(3), 1993 (Oct.); pp. 42-46.

Jodha, N.S. (1989), Technology Options and Economic Policy for Dryland Agriculture; Concept Publishing Company, New Delhi, 1989.

Johl, S.S. (1989), Role of Market Clearance in Promoting Diversified Agriculture Growth in India; Presidential Address; Indian Society of Agricultural Marketing held at Ahmedabad, Sept. 11-13, 1989.

Justin Yifn Lin (1988), The Household Responsibility system in China's Rural Reform, in Proceedings of the International Conference of Agricultural Economists, August 24-31, 1988, Buenos Aires, Argentina.

Kabir, Humayun (1992), Contribution of Factors to Sugarcane Productivity Differential Between Large and Small Farms in an Area of Bangladesh; *Bangladesh Journal of Agricultural Economics*, 15(2), 1992 (December); pp. 81-89.

Kailas Sarup (1991), Interlinked Agrarian Markets in Rural India; Sage Publishing Company, New Delhi, 1991.

Kainth, G.S. (1992), Strategy to Increase Sorgum Production; *Farmer and Parliament*, 27(8), 1992 (Aug.); pp. 25-29.

Kaneda, Hiromitsu (1980), Structural Change and Policy Response in Japanese Agriculture After the Land Reform; *Economic Development and Cultural Change*, Vol. 28(3), 1980.

Karna, M.N. (1981), Studies in Bihar's Economy and Society; Concept Publishing Company, New Delhi, 1981.

Khusro, A.M. (1973), Economics of Land Reforms and Farm Size in India; Macmillan Company Ltd., New Delhi, 1973.

Konar, Dhirendra Nath (1994), Progress of Agriculture in West Bengal in the Eighties; *Economic Affairs*, 39(1), 1994 (Jan.-March); pp. 54-60.

Krishnaji, N. (1984), Agrarian Relations and the Small Peasant *in* A.P. Economic Association Conference Papers, March, 1984.

Krishnaji, N. (1991), *Agricultural Price Policy in Asia*; IJAE, April-June, 1991, pp. 186-192.

Kurien, V. (1992), Agricultural and Rural Development in the 1990s and Beyond : What should India do and why?; *Artha Vijnana*, 34(1), 1992 (March), pp. 1-11.

Lal, Manohar and Das, Y.L. (1993), Impact of Inftrastructure on Agricultural Development in Bihar : A Case of Rural Roads; *Bihar Journal of Agricultural Marketing*, 1(4), 1993 (Oct.-Dec.); pp. 450-456.

Lardy, Nicholas R. (1986), Prospects and Some Policy Problems of Agricultural Development in China; *American Journal of Agricultural Economics*, 68(2), 1986.

Lau, L.J. and P.A. Yotopolous (1971), A Test for Relative Efficiency and Application to Indian Agriculture; *American Economic Review*, 1971.

Leibenstien, H.; Allocative Efficiency Vs. X-efficiency; *American Economic Review*, Vol. 61, pp. 392-415.

Lewis, W.A. (1955), The Theory of Economic Growth, London, George Allen & Unwin, 1955.

Lipton, Michael (1989), New Strategies and Successful Examples for Sustainable Development in the Third World; IFPRI, 1776, M. Avenue, N.W., Washington D.C. (No. 170) June, 1989.

Lipton, Michael; Land Assets and Rural Poverty, Staff Working Paper, No. 744, IDS, Sussex.

Mahajan, V.S. (1989), Studies in Indian Agriculture and Rural Development; DEEP & DEEP Publications, New Delhi, 1989.

Malik, R.S. and Kuhad, M.S. (1993), Efficient Water Management for Vegetable Production—Why and How?; *Farmer and Parliament*, 28(11), 1993 (Nov.); 23 and 25.

Mandal, R.B.; Land Utilization; Concept Publishing Company, New Delhi.

Maiti, R.G. (1993), Turning Drought Areas Fruitful; *Intensive Agriculture*, 30(11-12), 1993 (Jan.-Feb.); pp. 14-15.

Maji, C.C. and T. Haque (1990), Economics of Marginal and Small Farms in West Bengal—Paper presented at the National Seminar on Economics of Agriculture with Special reference to viability and sustainability of small and marginal farmers held at NIRD, Hyderabad, Oct. 8-10, 1990.

Maji, C.C., T. Haque and A. Bhattacharya, Small Farms and Surplus Generation—A Case of West Bengal. NCAP, Policy Paper No. 5, 1995.

Majumdar, Dipak (1963), On the Economics of Relative Efficiency of Small Farmers; *Economic Weekly*, Special Number, 1963.

Marshall, Alfred (1956), Principles of Economics, 8th edition, London, Macmillan, 1956.

Mellor, J.W. and R. Ahmed. Agricultural Price Policy for Developing Countries, John Hopkins Press, 1988.

Mellor, J.W. (1986), Agriculture on the Road to Industrialisation, IFPRI, Washington, D.C. Reprint No. 83, 1986.

Mathur, D.P. (1979), Management in Small Poultry Farms : A Study in Maharashtra and Gujarat Regions by D.P. Mathur and V.K. Gupta, New Delhi, Abhinav, 1979, 128p.

Misra, D.C. (1993), Agricultural Extension for Irrigated Commands in India; *Agricultural Situation in India*, 48(4), 1993 (July), pp. 231-43.

Mudgal, V.D. (1987), Buffaloes as the Backbone of Small Farms Economy in India; *National Bank News Review*, 3(10), 1987 (December); pp. 37-39.

Murray, J. Leaf (1994), Indian Agriculture and Agrarian Communities : Two Studies; *Economic Development and Cultural Change*, 42(3), pp. 647-53; April, 1994.

Nadkarni, M.V. (1993), Agricultural Policy and Ecology; *Social Action*, 43(3), 1993 (July-Sept.); pp. 271-80.

Nageswar Rao, A. and Dastagiri, M.D. (1988), Comparative Economics of Sheep Farming of Small Loan and Non-Loan Farms; *Financing Agriculture*, 20(2), 1988 (April-June), pp. 24-29.

National Bank News Review (1992), Evaluation of National Watershed Development Programme of Special Assistance to Small and Marginal Farmers, 8(5), pp. 32-34.

Newbery, D.M.G. (1977), Risk Sharing, Share Cropping and Uncertain Labour Markets; *Review of Economic Studies*, Vol. 44, Oct., 1977.

Newbery, D.M.G. (1974), Crop Sharing Tenancy in Agriculture—Comment; *American Economic Review*, Vol. 64, Dec. 1974, pp. 1060-80.

Newman, David H. and Wear David N. (1993), Production Economics of Private Forestry : A Comparison of Industrial and Non-industrial Forest Owners; *American Journal of Agricultural Economics*, 75(3), 1993 (August); pp. 764-84.

Nurul Islam (1990), Horticultural Exports of Developing Countries; Past Performances, Future Prospects and Policy Issues; IFPRI, Research Report No. 80, 1990.

Onte, Renato P. (1992), Productivity and Employment Implications of Small Scale Farming in Philippines, (CIRDAP Study series No. 101), Dhaka, 1992.

Pablo T. Tamesis (1992), Organisation and Management of Agricultural Services and Inputs to Small Farmers, Paper presented at the Regional Workshop held at CIRDAP, Dhaka, June 28-July 2, 1992.

Palakudy, Thomas and Singh, Sukhpal (1993), The New Agricultural Policy and the Rural Poor; *Social Action*, 43(3), 1993 (July-Sept.); pp. 328-38.

Pal, D.P. and Mondal, S.K. (1994), The Changing Pattern of Land Holdings in Rural India; *Economic Affairs*, 39(1), 1994 (Jan.-March); pp. 41-53.

Pal, Sasanka, "Impact of Tenancy Reforms on Production and Income Distribution," NCAP Policy Paper No. 1, 1995.

Patel, A.R. (1987), Financing Beneficiaries of Land Reform Policy; *Kurukshetra*, 35(9), 1987 (June); pp. 27-30.

Patel, G.N. and Others (1988). Credit Requirements and Gap on Small and Marginal Bidi Tobacco Farms in Central Gujarat; *Indian Cooperative Review*, 26(2), 1988, (October); pp. 121-27.

Pathak, B.S. (1993), Agro-Exports : Problems and Prospects; *Financing Agriculture*, 25(2), 1993 (April-June); pp. 22-27.

Pathak, R.K. (1993), Horticulture in Wasteland Development; *Financing Agriculture*, 25(2), 1993 (April-June); p. 12-18.

Pathak, P.A. (1993), Development of Non-farm Activities Through Institutional Credit; *National Bank News Review*, 8(11-12); Jan.-Feb., 1993, pp. 33-35.

Patnaik, Prabhat (1994), International Capital and National Economic Policy : A Critique of India's Economic Reforms; *Economic and Political Weekly*, 29(12), 1994 (March 1-9); pp. 683-89.

Patnaik, Utsa (1987), Peasant Class Differentiation; Oxford University Press, Delhi, 1987.

Parthasarathy, G. (1987), Changes in the Incidence of Rural Poverty and Recent Trends in Some Aspects of Agrarian Economy, *Indian Journal of Agricultural Economics*, Vol. XXXXII, Jan.-March, 1987, pp. 1-21.

Pawar, Rajinderjit and Singh, Raghbir (1993), Working and Living Conditions of Landless Agricultural Labourers of Punjab : An Empirical Study; *Indian Journal of Labour Economics*, 36(2), 1993 (April-June); pp. 224-32.

Prasad Arbind (1993), Emerging Trends in Agricultural Marketing : Some Policy Implications; *Bihar Journal of Agricultural Marketing*, 1(4), 1993 (Oct.-Dec.); pp. 367-73.

Prasad, Jagdish (1993), Dunkel Proposals and Indian Agriculture : A Synoptic View; *Bihar Journal of Agricultural Marketing*, 1(4), 1993, (Oct.-Dec.); pp. 398-403.

Prasad, Rajendra (1993), Making Rural Market an Effective Instrument for Small Farms Development; *Bihar Journal of Agricultural Marketing*, 1(1), 1993, (Jan.-March); pp. 14-26.

Pretty, Jules (1994), Alternative Systems of Inquiry for a Sustainable Agriculture; *IDS Bulletin*, 25(2), 1994 (April); pp. 37-49.

Radhamohan Mallik (1994), Farm Size-productivity Nexus : An Economic Analysis; *Man & Development*, 16(1), 65-86, March, 1994.

Rai, Jaspal (1993), Floriculture : Export Potential to Japan; *Rural Banker*, 2, 1993 (Oct.-Dec.); pp. 20-21.

Rajagopalan, V. (1994), Towards Development of Subsistance Farmers : The Need for New Approaches for Strategies; *Indian Journal of Agricultural Economics*, 49(1), Jan.-March, 1994; pp. 1-33.

Raj, Krishna (1964), Some Production Functions for the Punjab; *Indian Journal of Agricultural Economics*, Vol. XIX, No. 3&4, July-December, 1964, pp. 87-97.

Ramana, D.V. (1987), Economics of Sericulture and Silk Industry in India; Deep & Deep Publications, New Delhi, 1987.

Ramaswamy, C. and Otsuka, Keijiro (1992), Differential Adoption of Modern Rice Technology and Labour Market Adjustments in South India; *Bangladesh Development Studies*, 20(1), 1992 (March); pp. 93-107.

Rao, C.H. Hanumantha (1975), Technological Change and Distribution of Gains in Indian Agriculture, New Delhi, 1975.

Rao, C.H.H. Technological Change in Indian Agriculture : Emerging Trends and Prospects; *Indian Journal of Agricultural Economics*, Oct.-Dec., 1989, pp. 385-98.

Rao, C.H.H. (1971), Uncertainty Entrepreneurship and Share Cropping in India; *Journal of Political Economy*, Vol. 79, May-June, 1971, pp. 578-95.

Rao, C.H. Hanumantha (1966), An Alternative Explanations of the Inverse Relationship between Farm Size and output per Acre in India; *Indian Economic Review*, Vol. I, 1966.

Rao, R.S. and Tripathy, P.K. (1993), Identification of Marginal and Small Farmers—Their Numbers and Characteristics in the State of Orissa; *Artha Vijnana*, 35(3), 1993 (Sept.); pp. 230-40.

Rawal, T. (1992), Productivity and Employment Implications of Small Farming : Nepal; CIRDAP Study Series No. 100, Dhaka, 1992.

Reddy, K.D. (1993), Drip Irrigation; *Rural Banker*, 2, 1993, (Oct.-Dec.): pp. 16-19.

Reid, J.D. Jr. (1975), Share Cropping in History and Theory Agricultural History, Vol. 49, 1975, pp. 426-40.

Rayudu, C.S. (1991), Agricultural Credit and Rural Development in Drought Regions : A Study of Cooperative Banks of A.P.; Concept Publishing Company, New Delhi, 1991.

Riad, M., E.L. (1984), Development Strategy for the Rural Poor : Analysis of Country Experiences in the Implementation of WCAARD (Chapter III) Ghonemy, Rome, 1984.

Rodge, R.P. and Others (1992), Cropping—System Planning for Semi-arid Marathwada Region under Limited Availability of Irrigation Water; *Indian Journal of Agricultural Science*, 62(7), 1992 (July); pp. 436-41.

Rosegrant, Mark W. and R.E. Evenson (1993), Agricultural Productivity Growth in Pakistan and India; *Pakistan Development Review*, 32(4), 1993, pp. 443-48.

Roy, Indrajit and Others (1993), Transfer of Technology System in Bangladesh Agriculture; *Asia Pacific Journal of Rural Development*, 3(1), 1993 (July); pp. 64-70.

Rudra, Ashok (1978), Farm Size and Yield Per Acre; *Economic and Political Weekly*, 1978, pp. 26-28.

Rutten, Mario (1991), Capitalist Entrepreneurs and Economic Diversification; Academisch Redactic Bureau, Netherlands.

Sawant, S.D. (1990), Distribution of Land Holdings from the National Sample Survey; *Indian Journal of Agricultural Economics*, April-June, 1990, pp. 134-40.

Saikia, Ashok (1994), Evolution of Agricultural Policy in India; *Administrator*, 39(1), 1994 (Jan.-March), pp. 47-56.

Salahuddin Ahmed and Shafiqur Rahman (1992), Small Farmers' Use and Attitudes to the Delivery System, Paper presented at the Regional Workshop held at Dhaka, June 28-July 2, 1992.

Salahuddin Ahmed and Shafiqur Rahman (1992), Subsistence Farming in Asia : Income and Resource Allocation; *Asia-Pacific Journal of Rural Development*, 2(2), December, 1992, pp. 36-76.

Sangwan, P.S. (1994), Bio-fertilizers : Boon for Poor Farmers; *Rural India*, 57(1), 1994 (Jan.); pp. 20-21.

Sen, Bandhudas (1967), Farm Productivity and Soil Fertility in Indian Agriculture; *Indian Journal of Agricultural Economics*, Vol. XXII, April-June, 1967, No. 2.

Sen, A.K. (1962), An Aspect of Indian Agriculture; *The Economic Weekly*, Vol. XIV, Nos. 4, 5&6, Feb., 1962.

Sen, S.R. (1985), A Strategy for Our Small Farms; *Indian Economic Review*, 20(1), 1985 (Jan.-June); pp. 143-58.

Sengupta, Nirmal (1993), User Friendly Irrigation Design, MIDS; Sage Publishing Company, New Delhi, 1993.

Sengupta, Nirmal (1991), Managing Common Property; Sage Publishing Company, New Delhi, 1991.

Shah, C.H., Vimal Shah and Sudershan Iyengar (1991), Agricultural Growth and Equity; Concept Publishing Company, New Delhi, 1991.

Shanmugam, T.R. and Ramaswami, C. (1993), Milk Production in India—Statewise Analysis; *Economic Affairs*, 38(3), 1993 (July-Sept.), pp. 189-92.

Sharma, S.C. (1991), Utilisation of Wastelands for Sustainable Development in India; Concept Publishing Company, New Delhi, 1991.

Sharma, B.R. and Ramesh Chand (1993), Diversification of Agriculture : An Aid to Employment Generation in India; *National Bank News Review*, 8(11-12), Jan.-Feb., 1993, pp. 28-32.

Sharma, R.K. (1992), Role of Farm Extension in Drought Management; *Agricultural Situation in India*, 46(12), 1992 (March), pp. 935-37.

Sharma, D.V.S.R. (1993), Retrospect and Prospects of Introduction of Sprinkler System of Irrigation in Command Areas; *Bhagirath*, 40(2), 1993 (April-June), pp. 59-61.

Sharma, R.A. (1993), The Land-use Economy of Orissa with Special Reference to Forestry; *Indian Forester*, 119(7), 1993 (July), pp. 536-43.

Sharma, R.S. and Kewat, M.L. (1993), Lehi Method of Rice Cultivation does well in Madhya Pradesh; *Indian Farming*, 43(6), 1993 (Sept.), pp. 3-4.

Sharma, V.K. (1993), Future Prospects of Horticulture in Himachal Pradesh; *Farmer and Parliament*, 28(10), 1993 (Oct.), pp. 25-27.

Sidhu, D.S. (1990), Some Aspects of Agricultural Marketing and Pricing Policies in India; IJAE, Oct.-Dec., 1990, pp. 433-44.

Singh, B.P. and Ghosh, D.C. (1992), Effect of Cultural Practices on Productivity and Economics of Rice (Orizasativa) Wheat (Triticum aestivum) Croping System in Plateau Region of Bihar; *Indian Journal of Agricultural Science*, 62(1), 1992 (Jan.), pp. 20-23.

Singh, Inderjit (1988), Small Farmers in South Asia : Their Character in Productivity and Efficiency; Discussion Paper No. 31, 1988.

Singh, I.J. (1980), Agricultural Instability and Farm Poverty in India; *Indian Journal of Agricultural Economics*, Jan.-March, 1989, pp. 1-16.

Singh, Karam (1992), Case Studies of Successful Farmers Societies and Private Entrepreneurs : Experiences and issues; *Indian Journal of Agricultural Economics*, 1992, pp. 347-54.

Singh, Gian (1988), Rural Youth and Small Farms; *Intensive Agriculture*, 26(9), 1988 (Nov.), pp. 14-15.

Singh, Rachhpal and Sharma, A.C. (1987), Risk Efficient Production Planning on Small Farms in the Rainfed Sub-mountainous areas of Roopnagar District of the Punjab State; *Economic Affairs*, 32(2), 1987 (June), pp. 89-93.

Singh, Rachhpal and Sharma, A.C. (1988), Farm Production Plans for Risk Averters : A Case Study of Small Peasant Farms; *Indian Journal of Economics*, 69(272), 1988 (July), pp. 59-70.

Singh, Ranveer and Bhati, J.P. (1992), Increasing Farm Income and Employment in Hilly Areas : A Study in Himachal Pradesh; *Agricultural Situation India*, 46(12), 1992 (March), pp. 921-24.

Singh, S.P. (1993), Management Model for Agricultural Development : Problems and Perspectives; *Indian Journal of Public Administration*, 39(4), 1993 (Oct.-Dec.); pp. 677-84.

Sivasubramaniyan, K. (1994), Towards Revival of Small Water Bodies; *Economic and Political Weekly*, 29(30), July 23, 1994, 1936-37.

Singh, Tarlok, T. Haque and A.V.S. Reddy. Impact of Land Reforms on Agriculture and Rural Development, NIRD, Hyderabad, 1992.

Singh, Tarlok (1969), Poverty and Social Change with a Reappraisal: Orient Longman, Bombay, 1969.

Somani, L.L. Bio-Fertilisers in Indian Agriculture; Concept Publishing Company, New Delhi

Srinivasa Gowda, M.V. and Others (1989), Productivity Difference Between Small and Large Farms : An Econometric Evidence; *Agricultural Situation in India*, 42(10), 1988 (January); pp. 903-907.

Srivastava, Alka and Prasad, B. (1989), A Study of Income, Savings and Investment on Small, Medium and Large Farms of Rudrapur Block of District Nainital; *Agricultural Situation in India*, 44(1), 1989 (April), pp. 9-12.

Srivastava, L.S. (1993), Manage Drylands to Sustain Agriculture; *Financing Agriculture*, 25(2), 1993 (April-June), pp. 19-21.

Swaminathan, M.S. (1991), Science and Integrated Rural Development; Concept Publishing Company, New Delhi, 1991.

Tanaka, Koji (1993), Farmers Perceptions of Rice Growing Techniques in Laoa : "Primitive" or "Thammasat"?; South-East Asian Studies, 31(2), 1993 (Sept.), pp. 132-40.

Taslim, M.A. (1993), Redistributive Land and Tenancy Reform in Bangladesh Agriculture; *Journal of Developing Areas*, 27(3), 1993 (April), pp. 341-76.

Tewari, S.C. and Others (1987), Economic Status of Marginal and Small Farms : A Case Study of Kawal Catchment Area of Himachal

Pradesh; *Agricultural Situation in India*, 42(9), 1987 (December), pp. 793-99.

Tendler, J. (1993), Tales of Dissemination in Small-farm Agriculture : Lessons for Institution Builders; World Development, 21(10), 1993 (Oct.), pp. 1567-82.

Thakur, Yas Pal and Others (1993), Problems and Prospects of Broiler Farming in Himachal Pradesh; *Indian Farming*, 43(5), 1993 (Aug.), pp. 11-16.

Tomita, Masahiko (1991), Research on the Integrated System of Rain-fed Rice Cultivation Area in South Asia; Utsunomiya University, March 1991.

Tschiersch, Joachim (1990), Agricultural Policy and Policy Failure in Developing Countries; *Economics*, 42, 1990, pp. 119-27.

Tyagi, D.S. (1990), Evolution of Agricultural Prices Policy in India; Quarterly Economic Report, 33(2), 1990 (June), pp. 15-20.

Uma Lele and Manmohan Agarwal (1989), Small-Holder and Large-Scale Agriculture : Are there Trade offs in Growth and Equity; Media Discussion Paper No. 6, 1989.

Umesh, K.B. and Bisaliah, S. (1986), Productivity Differential between Small and Large Farms : An Econometric Study; *Asian Economic Review*, 28(3), 1986 (Dec.); pp. 17-33.

Uwe Otzen (1993). Reflections on the Principles of Sustainable Agricultural Development; *Environmental Conservation*, 20(4), 1993 (Winter), pp. 310-16.

Vaidya, C.S. (1993). Economic Viability of Marginal and Small Farms in Hill Agriculture of Himachal Pradesh; *Bihar Journal of Agricultural Marketing*, 1(4), 1993 (Oct.-Dec.), pp. 440-49.

Vyas, V.S. (1976), Structural Changes in Agriculture and the Small Farmers Sector, *Economic and Political Weekly*, Jan. 10, 1976.

Vyas, V.S. (1987), India's Rural Development Strategies : Lessons in Agricultural Growth and Poverty Alleviation in China's Rural Development Miracle by J.W. Longworth (ed.), IAAE, 1987.

Walker, T.S. and James J. Ryan. Village and Household Economics in India's Semi-Arid Tropics, Johns Hopkins University Press, 1990.

World Bank (1987), EDI, Seminar on Agricultural Marketing Policy—Country Papers, Washington, 1987.

Yadav, J.S. (1991), Evaluation of Agricultural Extension : A Study of Haryana; Concept Publishing Company, New Delhi, 1991.

Yadav, I.P.S. and Others. Comparative Study of Management of Various Farm Inputs for Berseem Production Between Small and Large Farms Around Jhansi District; *Indian Dairyman*, 41(11), 1989 (Nov.), pp. 578-81.

Yugandhar. B.N. and Neela Mukherjee. Studies in Village India; Concept Publishing Company, New Delhi, 1991.

Zerkovic, Milicia (1987), Issues in Indian Agricultural Development, West View Press, 1987.

Zarkovic, Milicia (1989). Linkages with the Global Economy : The Case of Indian and Yugoslav Agricultural Regions, *Peasant Studies*, Vol. 10, No. 3, 1989.

INDEX